FARM ANIMAL MANAGEMENT

FARM ANIMAL MANAGEMENT
Principles and Practices

A. CHANDRASEKAR

KNOWLEDGE
BAKERS

KNOWLEDGE BAKERS

ISBN: 978-93-49964-97-6

Published by : **Knowledge Bakers**
Pune, India
Email: knowledgebakers2019@gmail.com

Preface

Farm animal management is a critical component of agricultural systems worldwide, as it directly impacts the productivity, health, and welfare of animals while contributing to food security and the economy. Effective management of farm animals requires a comprehensive understanding of animal behavior, nutrition, breeding, health, and environmental factors. In an era where sustainable practices, ethical considerations, and efficiency are increasingly emphasized, the role of farm animal management has become more complex and vital than ever before.

This book, titled Farm Animal Management: Principles and Practices, aims to provide a solid foundation for students, farmers, livestock managers, and professionals in animal husbandry. It offers practical insights into the essential principles and practices of managing farm animals in a way that optimizes both animal welfare and farm productivity.

The book begins by exploring the fundamental concepts of farm animal management, such as species selection, housing design, and feeding practices. It addresses the biological and behavioral needs of different farm animal species, including cattle, poultry, sheep, goats, pigs, and others, and how these needs are integrated into farm operations. The discussion extends to the role of breeding and genetic selection in improving animal performance and resistance to diseases.

In addition to the fundamental aspects of animal care, Farm Animal Management: Principles and Practices emphasizes the importance of disease prevention, health management, and biosecurity. The book provides practical guidance on vaccination schedules, parasite control, and strategies for maintaining a disease-free environment, crucial for the health and productivity of farm animals. Ethical considerations, such as animal welfare, humane handling, and the role of the farm manager in promoting ethical practices, are also thoroughly discussed.

The book delves into the economic aspects of farm animal management, focusing on cost-effective practices that contribute to profitability without compromising animal welfare or environmental sustainability. Topics such as labor management, feeding systems, and waste management are explored, offering a holistic view of the farm animal management system. Case studies and examples from diverse agricultural systems provide readers with real-world insights into the challenges and solutions faced by farm managers.

As the agricultural industry faces global challenges such as climate change, food security, and sustainability, Farm Animal Management: Principles and Practices offers guidance on how to adapt management strategies to meet these challenges while ensuring the long-term health and productivity of farm animals.

It is our hope that this book will serve as a valuable resource for anyone involved in farm animal management, providing both practical knowledge and the theoretical understanding necessary to make informed decisions that benefit both animals and farmers alike.

A. Chandrasekar

Contents

1

Introduction

All living things are divided into seven categories using Latin terms that can be understood worldwide. They progressively get smaller and more numerous, and each contains less living things. Classification of living things is used to help identify different animals and to group them together with their relatives.

The first and largest category is the Kingdom. To date there are five kingdoms: Animalia, which is made up of animals; Plantae, which is made up of plants; Protista, which is made up of protists (single-celled creatures invisible to the human eye); Fungi, which is made up of mushrooms, mold, yeast, lichen, etc; and Monera, which is made up of the three types of bacteria.

The next category is the Phylum. There are several phyla within each kingdom. The phyla start to break the animals (or plants, fungi, etc) into smaller and more recognizable groups. The best known phylum is Chordata, which contains all animals with backbones (fish, birds, mammals, reptiles, amphibians). There is also Arthropoda (insects, spiders, crustaceans); Mollusca (snails, squid, clam); Annelida (segmented worms); Echinodermata (starfish, sea urchins) and many, many more.

The next category that makes up the phyla is the Class. The class breaks up animals into even more familiar groups. For example, the phylum Chordata is broken down into several classes, including Aves (birds), Reptilia (reptiles), Amphibia (amphibians), Mammalia (mammals) and several others.

The next category is the Order. Each class is made up of one or more orders. Mammalia can be broken down into Rodentia (mice, rats), Primates (Old- and New-World monkeys), Chiroptera (bats), Insectivora (shrews, moles), Carnivora (dogs, cats, weasels), Perissodactyla (horses, zebras), Artiodactyla (cows), Proboscidea (elephants) and many more.

Orders can then be broken down into Families. The order Carnivora can be broken down into Canidae (dogs), Felidae (cats), Ursidae (bears), Hyaenidae (hyaenas, aardwolves), Mustelidae (weasels, wolverines), and many more.

The next category is the Genus. The family Felidae, for example, can be broken down into Acinonyx (cheetah), Panthera (lion, tiger), Neofelis (clouded leopard) and Felis (domestic cats).

Finally, the genus is broken down into the Species. The genus Panthera can be broken down to include Panthera leo (lion) and Panthera tigris (tiger). Note that the genus is placed in front of the species.

To complicate things, each category can, in some cases, be broken down into sub-categories. A sub-category is made when there are some animals that are basically the same but live in different areas and behave differently from each other or have some general differences in the anatomy. A prime example is phylum Chordata, class Reptilia, order Squamata. The order Squamata consists of both lizards and snakes because they are basically the same. However, snakes have some physical differences, mainly the lack of legs (although some lizards also lack legs). This warrants the creation of two sub-orders: Serpentes (snakes) and Sauria (lizards). These are then broken down into their own families.

Another example is phylum Chordata, class Mammalia, order Proboscidea, family Elephantidae, genus Elephas and species maximus. Elephas maximus (Asian elephant) is broken down into four sub-species depending on where the sub-species lives and the physical description of it. One example is the Ceylon elephant (Elephas maximus maximus). Note that the sub-species follows the species and that they always receive lower-case letters.

It should be noted that most animals are given common names recognized outside of the scientific community and that these names may be different in each country. For example, in Canada a certain type of spider is known as the daddy long-legs and is known as the harvester spider in the US. Likewise a type of fly in the US is called the daddy long-legs and is known as the crane fly in Canada. However, the Latin term remains the same.

It should lastly be noted that the Latin names of animals are always changing. The Aleutian Canada goose has gone through ten name changes, starting off as the non-sub-species Anser leucopreius and ending with the sub-species Branta canadensis leucopareia. The name will most likely change again in the future if it is decided that the Aleutian Canada goose is an entirely separate species from the Canada goose or if there are no subspecies at all and just the Canada goose.

Definition

The exact definition of taxonomy varies from source to source, but the core of the discipline remains: the conception, naming, and classification of organism groups. The exact relationship of systematics and classification to taxonomy also varies because the usage of the terms in biology originated independently. As points of reference, recent definitions of taxonomy are presented below:

1. Theory and practice of grouping individuals into species, arranging species into larger groups, and giving those groups names, thus producing a classification;
2. A field of science (and major component of systematics) that encompasses description, identification, nomenclature, and classification; see
3. The science of classification, in biology the arrangement of organisms into a classification.
4. "The science of classification as applied to living organisms, including study of means of formation of species, etc."
5. "The analysis of an organism's characteristics for the purpose of classification"

The varied definitions either place taxonomy as a sub-area of systematics (definition 2), or appear to consider the two terms synonymous. There is some disagreement as to whether biological nomenclature is considered a part of taxonomy (definitions 1 and 2), or a part of systematics outside taxonomy. For example, the last definition is paired with the following definition of systematics that places nomenclature outside taxonomy:

Systematics: "The study of the identification, taxonomy and nomenclature of organisms, including the classification of living things with regard to their natural relationships and the study of variation and the evolution of taxa".

HISTORY OF TAXONOMY

Early Taxonomists

Taxonomy has been called "the world's oldest profession", and naming and classifying our surroundings has likely been taking place as long as mankind has been able to communicate. It would always have been important to know the names of poisonous and edible plants and animals in order to communicate this information to other members of the family or group.

In the East, one of the earliest recorded pharmacopoeias was written by Shen Nung, Emperor of China (c. 3000 BC). He wanted to spread information related to agriculture and medicine, and is said to have tasted hundreds of plants with the goal of learning their medicinal value. Records after this are difficult to interpret for some time, but medicinal plant illustrations show up in Egyptian wall paintings from c. 1500 BC. The paintings clearly show that these societies valued and communicated the uses of different species, and therefore had a basic taxonomy in place.

Aristotle to Pliny the Elder

Historical records show that informally classifying organisms took place at least back to the days of Aristotle (Greece, 384-322 BC), who was the first to begin to classify all living things. Some of the terms he gave to animals, such as "invertebrates" and "vertebrates" are still commonly used today. His student Theophrastus (Greece, 370-285 BC) carried on this tradition, and wrote a classification of 480 plants called Historia Plantarum. Again, several plant groups currently still recognized can be traced back to Theophrastus, such as Cornus, Crocus, and Narcissus. The next major turn-of-the-millennia era taxonomist came in the form of Pliny the Elder (Rome, 23-79 AD). His elaborate 160-volume work Naturalis Historia described many plants, and even gave many of them Latin binomial names.

Pre-Linnaean Taxonomists

It was not until c. 1500 years later that taxonomic works became ambitious enough to replace the ancient texts. This is often credited to the development of sophisticated optic lenses, which allowed for the morphology of organisms to be studied in much greater detail. One of the earliest authors to take advantage of this leap in technology was Andrea Cesalpino (Italy, 1519-1603), who is often referred to as "the first taxonomist". His magnum opus De Plantis came out in 1583, and described over 1500 plant species. Two large plant families that he first recognized are still in use today: the Asteraceae and Brassicaceae. Then in the seventeenth century John Ray (England, 1627-1705) wrote many important taxonomic works. Arguably his greatest accomplishment was Methodus Plantarum Nova (1682), where he published over 18,000 plant species. At the time his classifications were perhaps the most complex yet produced by any taxonomist, as he based his taxa on many combined characters. The next major taxonomic works were produced by Joseph Pitton de Tournefort (France, 1656-1708). His work from 1700, Institutiones Rei Herbariae, included over 9000 species in 698 genera, and directly influenced Linnaeus as it was the text he used as a young student.

The Linnaean era

The Swedish botanist Carl Linnaeus (1707-1778) ushered in a new era of taxonomy. With his major works Systema Naturae 1st Edition in 1735, Species Plantarum in 1753, and Systema Naturae 10th Edition, he revolutionized modern taxonomy. His works implemented a standardized binomial naming system for animal and plant species, which proved to be an elegant solution to a chaotic and disorganized taxonomic literature. As a result the Linnaean system was born, and is still used in essentially the same way today as it was in the eighteenth century. Currently, plant and animal taxonomists regard Linnaeus' work as the "starting point" for valid names (at 1753 and 1758 respectively). Names published before these dates are referred to as "pre-Linnaean", and not considered valid (with the exception of spiders published in Svenska Spindlar). Even taxonomic names published by Linnaeus himself before these dates are considered pre-Linnaean.

The word taxonomy was introduced in 1813 by Candolle, in his Théorie élémentaire de la botanique.

Phylogenetics and cladistics

Today, traditional rank-based biological classifications persist in a structure largely unchanged since the 1700s; however, how the relationships of these taxa are investigated has changed drastically in recent decades. It is now common for biologists to devise a classification based on the results of phylogenetic analysis using DNA sequence data, and taxa are typically required to be clades. Although phylogenetics itself is fundamental to modern-day systematics, its use for the description of new taxa, and for their placement within a classification scheme, is not required.

Application

Biological taxonomy is a sub-discipline of biology, and is generally practiced by biologists known as "taxonomists", though enthusiastic naturalists are also frequently involved in the publication of new taxa. The work carried out by taxonomists is crucial for the understanding of biology in general. Two fields of applied biology in which taxonomic work is of fundamental importance are the study of biodiversity and conservation. Without a working classification of the organisms in any given area, estimating the amount of diversity present is unrealistic, making informed conservation decisions impossible. As conservation becomes ever more politically important, taxonomic work impacts not only the scientific community, but society as a whole.

Classifying organisms

Biological classification is a critical step in the taxonomic process, as it informs the user as to what the relatives of the taxon are hypothesized to be. Although the discipline of taxonomy itself does not deal with the investigations of how taxa are related to one another, it does serve to communicate these results to the user. To do this, it uses taxonomic ranks, including, among others (in order from most inclusive to least inclusive): Domain, Kingdom, Phylum, Class, Order, Family, Genus, and Species.

Taxonomic descriptions

The 'definition' of a taxon is encapsulated by its description. There are no set rules governing the definition of taxa, but the naming and publication of new taxa is governed by sets of rules. In zoology, the nomenclature for the more commonly used ranks (superfamily to subspecies), is regulated by the International Code of Zoological Nomenclature (ICZN Code). In the fields of botany, phycology, and mycology, the naming of taxa is governed by the International Code of Nomenclature for algae, fungi, and plants (ICN).

The initial description of a taxon involves five main requirements:

1. The taxon must be given a name based on the 26 letters in the Latin alphabet (a binomial for new species, or uninomial for other ranks).
2. The name must be unique (i.e. not a homonym).
3. The description must be based on at least one name-bearing type specimen.
4. It should include reference to enough attributes to differentiate the taxon from other taxa (ICZN Code, Article 13.1.1; ICN, Article 38.2). Both codes deliberately separate defining the content of a taxon (its circumscription) from defining its name.
5. These first four requirements must be published in a work that is obtainable in numerous identical copies, as a permanent scientific record.

However, often much more information is included, like the geographic range of the taxon, ecological notes, chemistry, behavior, etc. How researchers arrive at their taxa varies; depending on the available data, and resources, methods vary from simple quantitative or qualitative comparisons of striking features, to elaborate computer analyses of large amounts of DNA sequence data.

Phenetics

In phenetics, also known as taximetrics, organisms are classified based on overall similarity, regardless of their phylogeny or evolutionary relationships. It results in a measure of evolutionary "distance" between taxa. Phenetic methods have become relatively rare in modern times, largely superseded by cladistic analyses, as phenetic methods do not distinguish plesiomorphic from apomorphic traits. However, certain phenetic methods, such as neighbor joining, have found their way into cladistics, as a reasonable approximation of phylogeny when more advanced methods (such as Bayesian inference) are too computationally expensive.

Databases

Modern taxonomy uses database technologies to search and catalog classifications and their documentation. While there is no commonly used database, there are comprehensive databases such as the Catalogue of Life, which attempts to list every documented species. The catalogue listed 1.4 million species for all kingdoms as of May 2012, claiming coverage of more than 74% of the estimated 1.9 million species known to modern science.

Domestication

Domestication (from Latin domesticus) is the process whereby a population of living organisms is changed at the genetic level, through generations of selective breeding, to accentuate traits that ultimately benefit humans. A usual by-product of domestication is the creation of a dependency in the domesticated organisms, so that they lose their ability to live in the wild. It differs from taming in that a change in the phenotypical expression and genotype of the animal occurs, whereas taming is simply the process by which animals become accustomed to human presence. In the Convention on Biological Diversity, a domesticated species is defined as a "species in which the evolutionary process has been influenced by humans to meet their needs." Therefore, a defining characteristic of domestication is artificial selection by humans. Humans have brought these populations under their control and care for a wide range of reasons: to produce food or valuable commodities (such as wool, cotton, or silk) and for types of work (such as transportation, protection, warfare), scientific research, or simply to enjoy as companions or ornaments.

Plants domesticated primarily for aesthetic enjoyment in and around the home are usually called house plants or ornamentals, while those domesticated for large-scale food production are generally called crops. A distinction can be made between those domesticated plants that have been deliberately altered or selected for special desirable characteristics (see cultigen) and those plants that are used for human benefit, but are essentially no different from the wild populations of the species. Animals domesticated for home companionship are usually called pets while those domesticated for food or work are called livestock or farm animals.

Background

Charles Darwin was the first to describe how domestication, selection and evolution are interlinked, and based on natural heritable variation among individual plants and animals. Today we know that such natural variation is caused by mutations in genes coding for these traits, and by new combinations of already existing genetic variation, based on earlier mutations. Darwin described how the process of domestication can involve both unconscious and methodical elements. Routine human interactions with animals and plants create selection pressures that cause adaptation to human presence, use or cultivation. Deliberate selective breeding has also been used to create desired changes, often after initial domestication. These two forces, unconscious natural selection and methodical selective breeding, may have both played roles in the processes of domestication throughout history. Both have been described from man's perspective as processes of artificial selection.

The domestication of wheat provides an example. Wild wheat falls to the ground to reseed itself when ripe, but domesticated wheat stays on the stem for easier harvesting. There is evidence that this change was possible because of a random mutation that happened in the wild populations at the beginning of wheat's cultivation. Wheat with this mutation was harvested more frequently and became the seed for the next crop. Therefore, without realizing, early farmers selected for this mutation, which may otherwise have died out. The result is domesticated wheat, which relies on farmers for its own reproduction and dissemination.

The domestication of dogs provides another example. It is speculated that tamer than average wolves, less wary of humans, selected themselves as dogs over many generations. These wolves were able to thrive by following humans to scavenge for food near camp fires and garbage dumps, which gave them an advantage over more shy individuals. Eventually a symbiotic relationship developed between people and these proto-dogs. The dogs fed on human food scraps, and humans found that dogs could warn them of approaching dangers, help with hunting, act as pets, provide warmth, or supplement their food supply. As this relationship progressed, humans eventually began to keep these self-tamed wolves and breed from them the types of dogs that we have today.

In recent times, selective breeding may best explain how continuing processes of domestication often work. Some of the best-known evidence of the power of selective breeding comes from an experiment by Russian scientist, Dmitri K. Belyaev, in the 1950s. His team spent many years breeding the Silver Fox (Vulpes vulpes) and selecting only those individuals that showed the least fear of humans. Eventually, Belyaev's team selected only those that showed the most positive response to humans. He ended up with a population of grey-coloured foxes whose behavior and appearance was significantly changed. They no longer showed any fear of humans and often wagged their tails and licked their human caretakers to show affection. These foxes had floppy ears, smaller skulls, rolled tails and other traits commonly found in dogs.

Despite the success of this experiment, it appears that selective breeding cannot always achieve domestication. Attempts to domesticate many kinds of wild animals have been unsuccessful. The zebra is one example. Despite the fact that four species of zebra can interbreed with and are part of the same genus as the horse and the donkey, attempts at domestication have failed. Factors such as temperament, social structure and ability to breed in captivity play a role in determining whether a species can be successfully domesticated. In human history to date, only a few species of large animal have been domesticated. In approximate order of their earliest domestication these are: dog, sheep, goat, pig, ox, yak, reindeer, water buffalo, horse, donkey, llama, alpaca, Bactrian camel and Arabian camel.

Animals

1. According to evolutionary biologist Jared Diamond, animal species must meet six criteria in order to be considered for domestication:Flexible diet - Creatures that are willing to consume a wide variety of food sources and can live off less cumulative food from the food pyramid (such as corn or wheat), particularly food that is not utilized by humans (such as grass and forage) are less expensive to keep in captivity. Carnivores by definition feed primarily or only on animal tissue, which requires the expenditure of many animals, though they may exploit sources of meat not utilized by humans, such as scraps and vermin.
2. Reasonably fast growth rate - Fast maturity rate compared to the human life span allows breeding intervention and makes the animal useful within an acceptable duration of caretaking. Large animals such as elephants require many years before they reach a useful size.
3. Ability to be bred in captivity - Creatures that are reluctant to breed when kept in captivity do not produce useful offspring, and instead are limited to capture in their wild state. Creatures such as the panda, antelope and giant forest hog are territorial when breeding and cannot be maintained in crowded enclosures in captivity.
4. Pleasant disposition - Large creatures that are aggressive toward humans are dangerous to keep in captivity. The African buffalo has an unpredictable nature and is highly dangerous to humans; similarly, although the American bison is raised in enclosed ranges in the Western United States, it is much too dangerous to be regarded as truly domesticated. Although similar to the domesticated pig in many ways, Africa's warthog and bushpig are also dangerous in captivity.
5. Temperament which makes it unlikely to panic - A creature with a nervous disposition is difficult to keep in captivity as it may attempt to flee whenever startled. The gazelle is very flighty and it has a powerful leap that allows it to escape an enclosed pen. Some animals, such as the domestic sheep, still have a strong tendency to panic when their flight zone is encroached upon. However, most sheep also show a flocking instinct, whereby they stay close together when pressed. Livestock with such an instinct may be herded by people and dogs.

6. Modifiable social hierarchy - Social creatures whose herds occupy overlapping ranges and recognize a hierarchy of dominance can be raised to recognize a human as the pack leader:
    1. tapirs and rhinoceroses are solitary and do not tolerate being penned with each other
    2. antelope and deer except for reindeer are territorial when breeding and live in herds only for the rest of the year
    3. bighorn sheep and peccaries have nonhierarchical herd structures and do not follow any definite leader: instead males fight continuously with each other for mating opportunities
    4. musk ox herds (although having a defined leader) maintain mutually exclusive territories and two herds will fight if kept together.

However, this list is of limited use because it fails to take into account the profound changes that domestication has on a species. While it is true that some animals, including parrots, whales, and most members of the Carnivora, retain their wild instincts even if born in captivity, some factors must be taken into consideration.

In particular, number (5) may not be a prerequisite for domestication, but rather a natural consequence of a species' having been domesticated. In other words, wild animals are naturally timid and flighty because they are constantly faced by predators; domestic animals do not need such a nervous disposition, as they are protected by their human owners. The same holds true for number (4) - aggressive temperament is an adaptation to the danger from predators. A Cape buffalo can kill even an attacking lion, but most modern large domestic animals were descendants of aggressive ancestors. The wild boar, ancestor of the domestic pig, is certainly renowned for its ferocity; other examples include the aurochs (ancestor of modern cattle), horse, Bactrian camels and yaks, all of which are no less dangerous than their undomesticated wild relatives such as zebras and buffalos. Others have argued that the difference lies in the ease with which breeding can improve the disposition of wild animals, a view supported by the failure to domesticate the kiang and onager. On the other hand for thousands of years humans have managed to tame dangerous species like the elephants, bears and cheetahs whose failed domestications had little to do with their aggressiveness.

Number (6), while it does apply to most domesticated species, also has exceptions, most notably in the domestic cat and ferret, which are both descended from strictly solitary wild ancestors but which tolerate and even seek out social interaction in their domestic forms. Feral domestic cats, for example, naturally form colonies around concentrated food sources and will even share prey and rear kittens communally, while wildcats remain solitary even in the presence of such food sources. Zoologist Marston Bates devoted a chapter on domestication in his 1960 book The Forest and the Sea, in which he talks a great deal about how domestication alters a species: Dispersal mechanisms tend to disappear for the reason stated above, and also because people provide transportation for them. Chickens have practically lost their ability to fly.

Similarly, domestic animals cease to have a definite mating season, and so the need to be territorial when mating loses its value; and if some of the males in a herd are castrated, the problem is reduced even further. What he says suggests that the process of domestication can itself make a creature domesticable. Besides, the first steps towards agriculture may have involved hunters keeping young animals, who are always more impressionable than the adults, after killing their mothers.

Another strong factor deciding whether a species will be considered for domestication is quite simply the availability of more suitable (or even better already domesticated) alternatives. For example a community that had been introduced to domestication by neighboring peoples will generally find it much more practical, economical and time saving to import already domesticated species than experiment with wild animals (even if they are of the same species). Generally speaking, the species of animals originally domesticated by early humans in the interconnected landmasses of Eurasia and Africa were far superior, both in working capacity and in food production, than the species found in the other continents, namely the Americas and Oceania.

HISTORY OF THE DOMESTICATION OF ANIMALS

Hunting species: 20,000-10,000 years ago

During the most recent ice age, from about 20,000 years ago (see Ice Ages), large mammals such as bison roam on the sub-arctic tundra of Europe and Asia. They are preyed upon by two groups of hunters, both much smaller and weaker than themselves - but both with a sufficiently developed social system to enable them to hunt and kill in packs.

These hunters are humans and wolves. The typical pack of wolves and of humans is surprisingly similar. It is family-based, led by a dominant male whose female partner is likely to have an authority second only to his. Members of the group are friendly to each other but deeply suspicious of outsiders. All members (not just the parents) are protective of the newly born and the young. Both species are good at interpreting the moods of others in the group, whether through facial expression or other forms of body language.

Legend acknowledges these shared characteristics in stories of children suckled by wolves. The other side of the same coin, in real life, means that wolf cubs adapt easily to life among humans.

For mutual benefit

Humans and wolves are competing for the same prey, but there are advantages for both in teaming up. For the wolf, human ingenuity and the use of weapons mean a share in a greater number of kills - and perhaps even an occasional taste of larger victims, such as mammoth. For humans, the wolf's speed and ferocity is equivalent to a new weapon.

The partnership is natural. So, undoubtedly, is how it first comes about. People love to nurture any abandoned young animal, and a wolf cub is well adapted to learn the rules of a hierarchical human society (in which its place will be low). From this partnership all dogs derive. Unbelievable though it seems, every single breed of dog is descended from wolves.

For a species to become domesticated, it must be willing to breed in man's company. 'Breed in captivity', the more usual phrase, implies a simple case of exploitation. The reality is more complex. In terms of survival, those species which have developed a relationship with man have far outstripped their wild cousins.

The most numerous large mammals, apart from humans, are cows, sheep, goats, pigs, horses and dogs. Domestic cats easily outnumber their wild equivalents, as do chickens and turkeys. The domestication of animals is based on an ancient contract, with benefits on both sides, between man and the ancestors of the breeds familiar to us today.

Dogs: from 12,000 years ago

The earliest known evidence of a domesticated dog is a jawbone found in a cave in Iraq and dated to about 12,000 years ago. It differs from a wolf in that it has been bred to have a smaller jaw and teeth. Selective breeding affects a species quite rapidly, and is a natural process for man to initiate - probably at first by accident rather than intention. A particular puppy in a litter is favoured because it has an attractive coat, barks well, is unusually friendly or obedient, noticeably large or small.

This is the dog which is kept and in its turn has puppies. Its desirable characteristics become perpetuated.

Images in Egyptian paintings, Assyrian sculptures and Roman mosaics reveal that by the time of these civilizations there are many different shapes and sizes of dog. To use the word 'breed' may be anachronistic, though there is evidence that a dog very like the present-day Pekingese (almost as far as one can get from a wolf) exists in China by the 1st century AD.

By that time Roman ladies also have lap dogs; their warmth is believed to be a cure for stomach ache. A Roman writer of the period gives similarly practical reasons for selecting the colour of a dog: shepherds' dogs should be white (to distinguish them from wolves in the dark) but a farmyard dog should have a black coat (to frighten thieves).

Sheep and goats, cattle and pigs: 9000-7000 BC

The first animals known to have been domesticated as a source of food are sheep in the Middle East. The proof is the high proportion of bones of one-year-old sheep discarded in a settlement at Shanidar, in what is now northern Iraq. Goats follow soon after, and these two become the standard animals of the nomadic pastoralists - tribes which move all year long with their flocks, guided by the availability of fresh grass.

Cattle and pigs, associated more with settled communities, are domesticated slightly later - but probably not long after 7000 BC. The ox may first have been bred by humans in western Asia. The pig is probably first domesticated in China.

The first reason for herding sheep and goats, or keeping cattle and pigs in the village, is to secure a regular supply of fresh meat. The hunter is dependent on the luck of the chase; if more animals are killed than can be immediately consumed, meals from the surplus will be increasingly unpleasant as the days go by. The herdsman, by contrast, has a living larder always to hand and a supply of dairy products as well.

These animals also provide for almost every other need of neolithic man. While they are alive, they produce dung to manure the crops. When they are dead, leather and wool for garments; horn and bone for sharp points, of needles or arrows; fat for tallow candles; hooves for glue.

Draught animals: from 4000 BC

Of the four basic farm animals, cattle represent the most significant development in village life. Not only does the cow provide much more milk than its own offspring require, but the brute strength of the ox is an unprecedented addition to man's muscle power.

From about 4000 BC oxen are harnessed and put to work. They drag sledges and, somewhat later, ploughs and wheeled wagons (an almost simultaneous innovation in the Middle East and in Europe). The plough immeasurably increases the crop of wheat or rice. The wagon enables it to be brought home from more distant fields.

India and southeast Asia use another version of the domesticated ox, well adapted to hot wet conditions - the water buffalo. Whether dragging a plough-like tool through a flooded field or hauling a cart on a dry track, the buffalo is ideally suited to the role of a farm animal in rice-growing areas. Like other members of the ox family, it also provides a good supply of milk.

The buffalo is first domesticated somewhere in the near-tropical regions of Asia. Precisely where or when is not known, but buffaloes feature as domestic animals on the seals of the Indus civilization.

Cats: from before 3000 BC

Apart from dogs, cats are the only domesticated animals to dwell indoors with humans. It is also the only one which is solitary in the wild, as opposed to living in packs, herds or flocks. As a result the cat has been able to take what it wants from man (food, shelter, play) and to pay its dues in return (pest control) without losing contact with its original identity.

Cats have remained closer than other domesticated animals to their wild cousins, partly because it is so difficult to control their breeding. And they are more able than any other to fend for themselves, in the country or even in a city, if human support is withdrawn.

It is not known when cats are first domesticated. But by the time of the earliest civilization they have already acquired in the human mind a characteristic which they have never lost - the quality of mystery.

In the temples of Egypt cats are sacred animals, and are mummified in their millions. In folk stories of all nations a cat is the natural companion for people who possess an alarming second sight, such as witches.

Horses: 3000 BC

Humans acquire their most important single ally from the animal kingdom when they domesticate the horse, in about 3000 BC.

Wild horses of various kinds have spread throughout most of the world by the time human history begins. Their bones feature among the remains of early human meals, and they appear in cave paintings with other animals of the chase. Some of their earliest fossil remains have been found in America, but after arriving across the Bering Land Bridge they become extinct in that continent. They are reintroduced by European colonists in the 16th century.

A natural habitat of the wild horse is the steppes of central Asia. Here, with its ability to move fast and far, it can gallop out of harm's way and make the most of scarce grazing. And here, some 5000 years ago, humans first capture, tame and breed the horse. The original purpose, as with cattle, is to acquire a reliable source of meat and subsequently milk. But then, in a crucial development, tribesmen discover that they have at their disposal a means of transport.

With a horse beneath him, man's ability to move is improved out of all recognition. The next comparable moment in the story of human speed does not arrive for another 5000 years - with steam trains.

The first domesticated horses are of a size which we would describe as ponies. Horses of this kind were still living in the wild in Mongolia until quite recent times. Discovered there in the 1870s, and named Przewalski's horse, they survive now only in zoos.

The entire range of horses known to us, from the mighty carthorse down to the smallest ponies, is the result of human breeding. Other wild breeds, now extinct, have been added to the stock. One such example is the tarpan, which was the native breed in Europe.

Asses: 3000 BC

At much the same time as the wild horse is being domesticated in the region of the Black Sea and the Caspian, its cousin the ass or donkey (a member of the same equus family) is tamed in Egypt. At this time the donkey appears to have roamed wild in northeast Africa and up through the Fertile Crescent into Mesopotamia.

So both horse and the ass, from north and from south, become available to two of the earliest civilizations - in Mesopotamia and Egypt.

Silk moths: 3000 BC

In China an indigenous silk moth is co-opted for man's purposes. Bombyx mori is still the only insect to have been fully domesticated (in the sense that, unlike the bee, it cannot live in the wild and is not known in a wild form). The silk moth has lost the power to fly; its caterpillar can find no mulberry leaves for itself. The species exists, and survives, only because humans like silk.

The earliest known silk from bombyx mori was found in a bamboo basket unearthed by archaeologists in China. Other pieces in the same basket were from wild silkworms. The fragments date from between 2850 and 2650 BC.

Camels: 3000-1500 BC

As beasts of burden and transport, camels occupy an important place alongside horses and donkeys. Two small members of the camel family, the llama and the alpaca of south America, are domesticated first - probably before 3000 BC. At that time both species appear to have been on the verge of extinction. Domestication by the American Indians saves them. Neither the llama nor the alpaca exists now in the wild.

The larger of the two, the llama, is primarily a beast of burden, while the shaggy alpaca is valuable for its wool. Neither animal is strong enough to pull a plough or drag a cart - two important steps in the story of civilization which are denied to the early Americans.

In the parched regions of north Africa and Asia two different species of camel become the most important beasts of burden - the single-humped Arabian camel (in north Africa, the Middle East, India) and the double-humped Bactrian camel (central Asia, Mongolia). Both are well adapted to desert conditions. They can derive water, when none is available elsewhere, from the fat stored in their humps.

It is probable that they are first domesticated in Arabia some time after 1500 BC. By about 1000 BC caravans of camels are bringing precious goods up the west coast of Arabia, linking India with the Mediterranean and Mesopotamia.

Poultry and pigeons: 2000 BC

The red jungle fowl, a member of the pheasant family, lives in the forests and bamboo jungles of India and southeast Asia. The male makes an impressive crowing sound and is dignified by a comb on his head and wattles under his beak. Jungle fowl of this kind are captured and kept for their eggs and their flesh by about 2000 BC in Asia. It is thought that all domestic poultry in the world today are descended from this one species.

At much the same period, in Egypt, pigeons are first persuaded to live and breed in the proximity of humans - again as a reliable source of protein. But some 3000 years later it is discovered that they have an extra and unusual talent. Some of them can be trained to fly home.

Elephants: 2000 BC

India is the region where elephants are first tamed, during the Indus civilization. The two species of elephant are at this time widespread - the Indian elephant throughout temperate Asia as far west as Syria, and the African elephant in regions north and south of the Sahara. (The mammoth has become extinct by the end of the last glacial period, about 10,000 years ago - partly through climatic changes and partly at the hands of human hunters.)

It is not known when elephants are first trained to take part in war, but by the 3rd century BC they are a valuable military force in both India and north Africa. An ability to learn tricks also makes the elephant a performing animal, popular in the arena of the Roman circus.

The honey of the bee: before the 6th century BC

No doubt hunter-gatherers, when they find the honey of bees in a hollow tree, often risk a sting for the pleasure of sweetness. The story of beekeeping can be described as the search for safer and more convenient ways of robbing a bees' nest.

The turning point in the domestication of the bee is the discovery that a swarm of bees can be coaxed into a specific nest - one designed by man for his own convenience in collecting the honey, and with it the useful substance of beeswax.

It is not known when the beehive is first developed, but the Greeks in classical times use a design which for centuries remains standard in much of Europe. Known as a skep, it is a dome constructed from a continuous coil of woven straw - looking much like an upturned basket. It stands on a wooden platform with a hole in, through which the bees enter.

The disadvantage of such a system is that the removal of the honey involves disturbing the nest of the bees. From the 17th century, when wooden hives come into use, extra chambers are added for the collection of honey. But the major improvement in beekeeping techniques is the achievement of a 19th-century clergyman, L.L. Langstroth.

Rabbits: from the 1st century BC

Since Roman times, if not before, people have encouraged rabbits to breed in captivity for the sake of their meat, and have then regretted doing so because of the animal's ability to burrow to freedom and eat the crops. The only safe place to keep rabbits is on an island. (Almost every island of the world has rabbits on, brought by humans to establish a living larder for passing ships.)

Rabbits are inaccessible in their burrows, so man domesticates a species of polecat (in the form of the ferret) to flush them out. As early as the 1st century AD Pliny describes the use of ferrets in the Balearic islands, as the inhabitants struggle to control the rabbits (see Pliny and the ferrets).

The turkey: from the 14th century

he turkey is indigenous to central and north America. It is kept as a domestic fowl by the Aztecs in Mexico from the 14th century, and no doubt has been domesticated considerably earlier by their predecessors.

Turkeys are brought to Europe by the Spanish in the early 16th century. They become popular throughout the continent, and in the 17th century are taken back across the Atlantic by English settlers. Interbred with wild turkeys of north America, they develop into the breeds known there today.

The bee space: 1851

An American clergyman, the Rev. L.L. Langstroth, puts to practical use the very precise habits of bees. Other beekeepers in the 19th century have begun putting into the hives removable frames, in which the bees will build their honeycombs, but the bees clog the spaces up with wax and honey. In 1851 Langstroth discovers that if the frames are an exact distance apart (a little more than a quarter of an inch), the workers will keep each honey-filled frame neat and separate.

A gap which is larger or smaller does not have the same effect. The 'bee space' becomes an established principle, and the Langstroth hive is still the type used by beekeepers today.

The ostrich: 19th - 20th century

In addition to the standard domesticated animals, many others have been kept or are now kept by humans for a wide range of purposes.

A good example is the ostrich. In the late 19th-century the fashion for ostrich feathers, in hats or fans, causes farmers to domesticate and breed this largest of birds. In the late 20th-century there is more interest in it for its meat.

The hamster: 1930

In the ongoing story of domestication, the remarkable case of the golden hamster is a good detail for a closing chapter. In 1930 a female hamster with twelve young is captured at Aleppo in Syria. Taken to the laboratory at the Hebrew university in Jerusalem, they are bred for use in experiments. Each female has several litters a year, so the numbers rapidly grow. Other laboratories secure their own supply from Jerusalem. Then somebody notices that this little animal - chubby, with puffed out cheeks and soft fur - has considerable charm as a pet for children.

Half a century on the golden hamster is a common domesticated animal. Every single hamster in capitivity descends from that first Syrian litter.

Degrees

The boundaries between surviving wild populations and domestic clades can be vague. A classification system that can help solve this confusion surrounding animal populations might be set up on a spectrum of increasing domestication:

- **Wild:** These populations experience their full life cycles without deliberate human intervention.
- **Raised in Captivity/Captured from Wild:** These populations are nurtured by humans but (except in zoos) not normally bred under human control. They remain as a group essentially indistinguishable in appearance or behaviour from their wild counterparts. Examples include Asian elephants, animals such as sloth bears and cobras used by showmen in India, and animals such as Asian black bears (farmed for their bile), and zoo animals, kept in captivity as examples of their species. (It should be noted that zoos and botanical gardens sometimes exhibit domesticated or feral animals and plants such as camels, mustangs, and some orchids)
- **Raised commercially (captive or semidomesticated):** These populations are ranched or farmed in large numbers for food, commodities, or the pet trade, commonly breed in captivity, but as a group are not substantially altered in appearance or behavior from their wild cousins. Examples include the ostrich, various deer, alligator, cricket, pearl oyster, raptors used in falconry and ball python. (These species are sometimes referred to as partially domesticated.)
- **Domesticated:** These populations are bred and raised under human control for many generations and are substantially altered as a group in appearance or behaviour. Examples include sweet potato, garlic, pigs, ferrets, turkeys, canaries, domestic pigeons, budgerigars, goldfish, koi carp, silkworms, dogs, cats, sheep, cattle, chickens, llamas, guinea pigs, laboratory mice, horses, goats and (silver) foxes.

This classification system does not account for several complicating factors: genetically modified organisms, feral populations, and hybridization. Many species that are farmed or ranched are now being genetically modified. This creates a unique category because it alters the organisms as a group but in ways unlike traditional domestication. Feral organisms are members of a population that was once raised under human control, but is now living and multiplying outside of human control. Examples include mustangs. Hybrids can be wild, domesticated, or both: a liger is a hybrid of two wild animals, a mule is a hybrid of two domesticated animals, and a beefalo is a cross between a wild and a domestic animal.A great difference exists between a tame animal and a domesticated animal. The term "domesticated" refers to an entire species or variety while the term "tame" can refer to just one individual within a species or variety. Humans have tamed many thousands of animals that have never been truly domesticated. These include the elephant, giraffes, and bears. There is debate over whether some species have been domesticated or just tamed. Some state that the elephant has been domesticated, while others argue the cat has never been. Dividing lines include whether a specimen born to wild parents would differ in appearance or behavior from one born to domesticated parents. For instance a dog is certainly domesticated because even a wolf (genetically share a common ancestor with all dogs) raised from a pup would be very different from a dog, in both appearance and behaviour. Similar problems of definition arise when domesticated cats go feral.

Limits

Selection of animals for visible "desirable" traits may make them unfit in other, unseen, ways. The consequences for the captive and domesticated animals were reduction in size, piebald color, shorter faces with smaller and fewer teeth, diminished horns, weak muscle ridges, and less genetic variability. Poor joint definition, late fusion of the limb bone epiphyses with the diaphyses, hair changes, greater fat accumulation, smaller brains, simplified behavior patterns, extended immaturity, and more pathology are a few of the defects of domestic animals. All of these changes have been documented in direct observations of the rat in the 19th century, by archaeological evidence, and confirmed by animal breeders in the 20th century.

One side effect of domestication has been zoonotic diseases. For example, cattle have given humanity various viral poxes, measles, and tuberculosis; pigs and ducks have given influenza; and horses have given the rhinoviruses. Humans share over sixty diseases with dogs. Many parasites also have their origins in domestic animals. The advent of domestication resulted in denser human populations which provided ripe conditions for pathogens to reproduce, mutate, spread, and eventually find a new host in humans.

Dates and places

Since the process of domestication inherently takes many generations over a long period of time, and the spread of breed and husbandry techniques is also slow, it is not meaningful to give a single "date of domestication". However, it is believed that the first attempt at domestication of both animals and plants were made in the Old World by peoples of the Mesolithic Period. The tribes that took part in hunting and gathering wild edible plants, started to make attempts to domesticate dogs, goats, and possibly sheep, which was as early as 9000 BC. However, it was not until the Neolithic Period that primitive agriculture appeared as a form of social activity, and domestication was well under way. The great majority of domesticated animals and plants that still serve humans were selected and developed during the Neolithic Period, a few other examples appeared later. The rabbit for example, was not domesticated until the Middle Ages, while the sugar beet came under cultivation as a sugar-yielding agricultural plant in the 19th century. As recently as the 20th century, mint became an object of agricultural production, and animal breeding programs to produce high-quality fur were started in the same time period.

The methods available to estimate domestication dates introduce further uncertainty, especially when domestication has occurred in the distant past. So the dates given here should be treated with caution; in some cases evidence is scanty and future discoveries may alter the dating significantly.

Dates and places of domestication are mainly estimated by archaeological methods, more precisely archaeozoology. These methods consist of excavating or studying the results of excavation in human prehistorical occupation sites. Animal remains are dated with archaeological methods, the species they belong to is determined, the age at death is also estimated, and if possible the form they had, that is to say a possible domestic form. Various other clues are taken advantage of, such as slaughter or cutting marks. The aim is to determine if they are game or raised animal, and more globally the nature of their relationship with humans. For example the skeleton of a cat found buried close to humans is a clue that it may have been a pet cat. The age structure of animal remains can also be a clue of husbandry, in which animals were killed at the optimal age.

New technologies and especially mitochondrial DNA, which are simple DNA found in the mitochondria that determine its function in the cell provide an alternative angle of investigation, and make it possible to reestimate the dates of domestication based on research into the genealogical tree of modern domestic animals.

It is admitted for several species that domestication occurred in several places distinctly. For example, research on mitochondrial DNA of the modern cattle Bos taurus supports the archaeological assertions of separate domestication events in Asia and Africa. This research also shows that Bos taurus and Bos indicus haplotypes are all descendants of the extinct wild ox Bos primigenius. However, this does not rule out later crossing inside a species; therefore it appears useless to look for a separate wild ancestor for each domestic breed.

The first animal to be domesticated appears to have been the dog, in the Upper Paleolithic era. This preceded the domestication of other species by several millennia. In the Neolithic a number of important species such as the goat, sheep, pig and cow were domesticated, as part of the spread of farming with characterises this period. The goat, sheep and pig in particular were domesticated independently in the Levant and Asia.

There is early evidence of beekeeping, in the form of rock paintings, dating to 13,000 BC.

Recent archaeological evidence from Cyprus indicates domestication of a type of cat by perhaps 9500 BC.

The earliest secure evidence of horse domestication, bit wear on horse molars at Dereivka in Ukraine, dates to around 4000 BC. The unequivocal date of domestication and use as a means of transport is at the Sintashta chariot burials in the southern Urals, c. 2000 BC. Local equivalents and smaller species were domesticated from the 26th century BC.

The availability of both domesticated vegetable and animal species increased suddenly following the voyages of Christopher Columbus and the contact between the Eastern and Western Hemispheres. This is part of what is referred to as the Columbian Exchange.

ANIMAL HUSBANDRY

Animal husbandry is the dynamic set of relationships between humans and animals, in which qualities considered to be advantageous to humans are emphasized and further developed. The term can refer to the practice of selectively breeding and raising livestock to promote desirable traits in animals for utility, sport, pleasure, or research, but also refers to the efficient exploitation of a species in agriculture in a way that benefits all species involved in the relationship.

History of breeding

Animal husbandry has been practiced for thousands of years, since the first domestication of animals. Selective breeding for desired traits was first established as a scientific practice by Robert Bakewell during the British Agricultural Revolution in the 18th century. One of his most important breeding programs was with sheep. Using native stock, he was able to quickly select for large, yet fine-boned sheep, with long, lustrous wool. The Lincoln Longwool was improved by Bakewell and in turn the Lincoln was used to develop the subsequent breed, named the New (or Dishley) Leicester. It was hornless and had a square, meaty body with straight top lines. These sheep were exported widely and have contributed to numerous modern breeds.

Under his influence, English farmers began to breed cattle for use primarily as beef for consumption - (previously, cattle were first and foremost bred for pulling ploughs as oxen). Long-horned heifers were crossed with the Westmoreland bull to eventually create the Dishley Longhorn. Over the following decades, farm animals increased dramatically in size and quality. In 1700, the average weight of a bull sold for slaughter was 370 pounds (168 kg). By 1786, that weight had more than doubled to 840 pounds (381 kg).

Animal herding professions specialized in the 19th century to include the cowboys of the United States and Canada, charros and vaqueros of Mexico, gauchos and huasos of South America, and the farmers and stockmen of Australia.

In more modern times herds are tended on horses, all-terrain vehicles, motorbikes, four-wheel drive vehicles, and helicopters, depending on the terrain and livestock concerned. Today, herd managers often oversee thousands of animals and many staff. Farms, stations and ranches may employ breeders, herd health specialists, feeders, and milkers to help care for the animals.

Breeding techniques

Techniques such as artificial insemination and embryo transfer are frequently used today, not only as methods to guarantee that females breed regularly but also to help improve herd genetics. This may be done by transplanting embryos from high-quality females into lower-quality surrogate mothers - freeing up the higher-quality mother to be reimpregnated. This practice vastly increases the number of offspring which may be produced by a small selection of the best quality parent animals. On the one hand, this improves the ability of the animals to convert feed to meat, milk, or fiber more efficiently, and improve the quality of the final product. On the other, it decreases genetic diversity, increasing the severity of disease outbreaks among other risks.

History in Europe

The semi-natural, unfertilized pastures formed by traditional agricultural methods in Europe, were managed and maintained by the grazing and mowing of livestock. Because the ecological impact of this land management strategy is similar to the impact of a natural disturbance, the agricultural system will share many beneficial characteristics with a natural habitat including the promotion of biodiversity. This strategy is declining in the European context due to the intensification of agriculture, and the mechanized chemical based methods that became popular during and following the industrial revolution.

Sustainability

In organic and biodynamic farming methods, animals play an important role in achieving closed or sustainable system by providing multiple functions to the agricultural system. In the Balinese context, for example, ducks are herded through rice paddies at particular stages of the growing season. The ducks eat insects and weeds that inhibit the growth of seedlings, and prevent the farmer from using herbicides or pesticides. Their droppings contribute nitrogen to the soil as a natural source of fertilizer, and the movement of their feet and bills in the water of the terrace ecosystem increases the oxygen content of the soil thus increasing nutrient availability.

Climate change

Due to the significant contribution of agriculture to the emissions of non-CO2 greenhouse gases methane and nitrous oxide, the relationship between humans and livestock is being analyzed for its potential to help mitigate climate change. Strategies for the mitigation include optimizing the use of gas produced from manure for energy production (biogas).

Animal husbandry in India

A large number of farmers in India depend on animal husbandry for their livelihood. In addition to supplying milk, meat, eggs,wool and hides, animals, mainly bullocks, are the major source of power for both farmers and drayers. Thus, animal husbandry plays an important role in the rural economy. The gross value of output from this sector was 358 billion (US$5.5 billion) in FY 1989, an amount that constituted about 25 percent of the total agricultural output of 1.4 trillion (US$21.4 billion).Production

In FY 1992, India had approximately 25 percent of the world's cattle, with a collective herd of 193 million head. India also had 110 million goats, 75 million water buffalo, 44 million sheep, and 10 million pigs. Milk production in FY 1990 was estimated to have reached 53.5 million tons, and egg production had reached a level of 23.3 billion eggs. Dairy farming provided supplementary employment and an additional source of income to many small and marginal farmers. The National Dairy Development Board was established in 1965 under the auspices of Operation Flood at Anand, in Gujarat, to promote, plan, and organize dairy development through cooperatives; to provide consultations; and to set up dairy plants, which were then turned over to the cooperatives. There were more than 63,000 Anand-style dairy cooperative societies with some 7.5 million members in the early 1990s. The milk produced and sold by these farmers brought 320 million (US$4.9 million) a day, or more than 10 trillion (US$153.0 billion) a year. The increase in milk production permitted India to end imports of powdered milk and milk-related products. In addition, 30,000 tons of powdered milk were exported annually to neighboring countries. There was about 50000 cows in India.Today, India has the world's largest dairy herd (composed of cows and buffaloes), at over 304 million strong, and stands first in milk production, with 112.5 million tonnes of milk produced in 2009-2010. India is also the third largest egg-producer in the world, at over 180 million eggs being produced every day or 65.7 billion eggs for the year 2011-12, and the world's sixth largest producer of poultry meat. While the majority of India's animal products are consumed domestically, exports are growing. India is the top global exporter of buffalo meat, and is also the fourth largest exporter of soybean meal, an important ingredient in commercial feed for farmed animals. In addition, India's leading poultry producers, including Suguna, Venky's, and the Amrit Group, are increasing sales to countries in other parts of Asia and the Middle East. International investment is also expanding. In 2008, U.S.-based Tyson Foods acquired a 51 percent stake in Godrej, an Indian conglomerate that is a major producer of animal feeds and poultry, and the International Finance Corporation (IFC) has an equity stake in Suguna. An estimated 200 million egg-laying hens are kept in battery cages, and more than half of the 2-billion-plus "meat" chickens produced each year are factory-farmed. Vertical integration and contract farming for poultry and eggs is also widespread. While the milk and cheese sectors still include many small-scale producers and cows and buffaloes in extensive systems the number of large, factory-style dairy operations is increasing.

Operation Flood

Operation Flood, the world's largest integrated dairy development program, attempted to establish linkages between rural milk producers and urban consumers by organizing farmer-owned and -managed dairy cooperative societies. In the early 1990s, the program was in its third phase and was receiving financial assistance from the World Bank and commodity assistance from the European Economic Community. At that time, India had more than 64,000 dairy cooperative societies, with close to 7.7 million members. These cooperatives established a daily processing capacity of 15.5 million liters of whole milk and 727 tons of milk powder.

PASTURE MANAGEMENT FOR DAIRY FARMERS

Good pasture management is one of the most important aspects of successful dairy production. Once an improved grass-legume pasture is established it can remain productive and weed free, providing large amounts of good quality feed for dairy cows, bulls, and young stock for many years. This will occur so long as the person managing the farm follows some simple rules, as follows:

Stocking Rate [SR]

Maintaining the appropriate stocking rate is perhaps the most important rule of pasture management. The right stocking rate will vary from farm to farm, depending on;

- Soil fertility
- Annual rainfall and its distribution
- Whether the pasture is shaded or unshaded
- Pasture variety (improved Vs unimproved)
- Use of fertiliser
- Pests and diseases of the pasture

Stocking rate is generally expressed in terms of animal units per hectare. One animal unit is equivalent to one cattle beast weighing 450 kg. Thus an animal of with a liveweight of 600 kg = 1.33 au, whilst one weighing 225 kg = 0.5 au. This is based on the assumption that the daily feed requirement of cattle is a constant percentage of their body weight: 3% of body weight in good quality pasture dry matter. This is not completely accurate as the percentage will vary for different ages of stock and quality of feed and level of production, never the less it is a very valuable yardstick which will allow a farmer or adviser to match demand and supply of feed in such a way as to achieve high desired sustainable production of milk and or beef.

As an example, the average the carrying capacity (SR) of improved (batiki + legume) pastures is;

- Open Pastures -2.5 au/ha
- Under coconuts -2.0 au/ha

These stocking rates represent the balance at which animal and plant [pasture] production can be sustained.

Assessing the appropriate level of stocking

The farm manager should monitor the feed supply (pasta re) availability of the farm at least twice a week. The purpose of this is to match feed supply (pasture growth) with the needs of the animals. Small fluctuations in feed supply due to short periods without rain can normally be accounted for by a little supplementary feeding with elephant grass or similar. If on the other hand pasture availability is too little or too much, changes in the way the farm is run should be made as soon as possible, as even though it may not be observable, production will be suffering.

If rotational grazing is used, the pasture should be about 30 cm high (mid-calf) when the cows start grazing and about 15 cm high (just above the ankle) when cows are removed. Under set stocking pasture should be of even height, about 15 - 20 cm.

PASTURE AT MID CALF HEIGHT PASTURE AT ABOVE ANKLE

Height

Other indicators of incorrect SR are;

1. More weeds observed in the pasture,- SR is too high.
2. Pasture height uneven , with tall clumps of grass, - SR is too low

What to do if SR is not appropriate

Too High?

This can easily happen, because the area grazed is limited but animals tend to grow and multiply.

Action to restore a good feed supply Should be taken urgently as if a milking cow 5 feeding is reduced for any length of time it is very difficult to make up losses in production later. The farmer has two alternatives,

1. Increase the feed supply by

a. Using more fertiliser, which should be spread evenly over the whole grazing area. Price and availability varies from country to country, for the most appropriate fertiliser, consult your local extension officer and Agricultural supplier. should be spread evenly over the whole grazing area. This should be repeated every six months. The stocking of the farm should probably not be raised higher than its current level.

b. Practising supplementary feeding or zero grazing, cut and carry grass and legumes are suitable in this case. To allow the pastures to grow back again, as much area as possible should be closed from grazing for about 3 weeks which given rain should allow it time to recover, after this it should be stocked at the appropriate stocking level by the milking cows. The balance of the stock should be held on the rest of the farm, as small an area as possible leaving as much as possible for milk production. Higher than normal levels of supplementary feeding to the dairy cows is also necessary.

2. Decrease the stocking rate, either by

a. Increasing the total grazed area, i.e. fencing more land or tethering some cattle outside the fenced area, such as roadsides and cropping land which have been harvested. This will effectively decrease the average stocking rate per hectare.

b. Sell some animals. This is a practice that all farmers must consider, particularly those with only small farms. To ensure the best animal health, production and profits from the farm animal numbers must be limited. This also adds to the farmers income through the sale of unneeded, old and unproductive animals.

The Farmer needs to answer these questions:

1. What is the main purpose of the farm?
2. Which cattle are necessary to obtain the objective.

Sample Answers

1. The main purpose in this example is to earn income from milk sales.
2. The necessary animals are:

heifer calves
yearling heifers
2 yr heifers
milking cows
bulls - (1 for every 30 cows)

Below is an example of an overstocked situation and how a farm manager can assess the ideal stocking rate and go about adjusting the herd to Suit.

Calculation of Farm Carrying Capacity

A small holder farm of 12 ha, of which:
9 ha pasture tinder 40 year old coconuts
3 ha open pasture
12ha 9@ 2au/ha = 18
3@ 2.5 au/ha = 7.5

The farmer should therefore consider the best way to reduce the number of animal units carried on the grazed area. (for simplicity we will not include using fertiliser in this example). The 2 main factors mentioned above should be applied in choosing which cattle should go (sold, tethered outside the fence, yard fed or transferred to another farm).

So a farm with about 25.5 au could be structured as follows:

This maximises the number of productive animals [milking cows] whilst balancing the stocking rate with the carrying capacity.

Actions taken to make the change were:

1. All old non productive cows sold
2. Sale of excess bulls
3. Sale of all bull calves as weaners
4. Sale of one weaner heifer (keep the best)
5. Sale of 3 yrs steers

Sale of 3 2 yrs steers
Sale of 5 mixed age steers

The following increases are made:

1. purchase of a 1 year bull (to avoid inbreeding)
2. Increase in milking cows
3. Increase in calves

By restructuring the herd, the farmer has simplified the operation and been able to increase the number of productive animals i.e. cows by 3.

This will increase the milk production. Production increases as a result of better feeding would be expected to be in the order of:

1. Decrease in calving interval from 540 to 410 days would increase annual calving percentage.
2. Milk production per lactation would increase by approx. 100% or more per cow.

A Note on Pasture Management

Good pasture management is more important in dairy farming than in beef production. This is because

1. Expenses (and returns) tend to be higher, therefore all aspects of management need to be good to combine to a high level of production.
2. Whilst beef animals are to some extent able to rely on compensatory growth to even out low production of beef caused by periods of poor or low quantity feeding, dairy cows cannot so easily recoup lost milk production, after poor feeding.

Good pasture management is: producing constant large quantities of good quality forage to be converted into useful products by grazing cattle year after year.

Grass-Legume Balance

The pasture should ideally contain at least 20% legume (leaf) on a DM (dry matter) basis High legume content increases the protein content and per cow intake of the pasture.

High legume content can be encouraged by:

1. planting recommended pasture legumes
2. good management of newly planted pastures/legumes encouraging their spread in the pasture, during the first 12 months especially.
3. strategic heavy grazing in the early wet season to encourage germination of aruinal legumes such as Glenn jointvetch,
4. the use of phosphate fertilisers.

Pasture Growth Stage

The grass tends to be the dominant plant in an improved pasture. Whilst tropical grasses are highly effective at photosynthesis and thus compete well for space and light they tend to be very low in crude protein. Tropical grasses tend to have decreasing levels of protein and as they mature (leaf age). The critical crude protein level is 7% by DM of the diet. Below this feed intake is depressed as the cow cant digest feed quickly enough in turn production is depressed. As an example, batiki is the major pasture grass in Samoa and tends to have a low CP% which falls quickly with leaf age (and increasing stem). Grasses should be kept young and leafy with a short grazing interval.

Rotational Grazing

This type of grazing will allow the farmer to graze the pasture and then leave it to recover. Portable electric fences are an effective, low cost method of controlling the grazing area and daily pasture allocation.

The length (in days) of the rotation may have to vary a little e.g. Shorter when the grass is tending to seed, this should help to promote vegetative growth.

If conditions are dry, the rotation may have to lengthen a little so that, when the cows enter a new paddock they are offered the same amount of feed i.e. 25-30 cm high. This will mean leaving them on the paddock a little longer and therefore the "residuals' (amount of pasture left) will be less than usual. This is only recommended for a short time and supplements should be increased in this case.

Worm burdens tend be less under rotational than continuous grazing.

The Use of Fertiliser

It is widely accepted internationally that to achieve high production levels over the long term, fertiliser inputs are necessary. The basic idea being that something is being removed from the agrisystem i.e. coconuts, meat, milk, - then something must be returned i.e. essential nutrients N,P,K,S (+ micro nutrients) otherwise the productive capacity (soil fertility) will decline overtime. Perhaps this is best described as land stewardship: ensuring that you are able to hand the land to your children in at least as good a condition as you received it from your parents.

Maintenance Fertiliser

The principle of maintaining the productive capacity of the land should be followed, fertiliser policies which farmers can apply for pasture development and maintenance in all cattle raising areas need to be prepared, your MAF may be able to advise you on this.

Strategic Use of Fertiliser

Fertiliser can be used to boost pasture plant growth in order to overcome (quickly) problems such as short term feed shortages and weed problems.

Banana fertiliser (NPK: 12-5-20) has been shown (Lee SD 1995) to be highly beneficial in controlling mintweed (H capucita) and navua sedge (K polyphylla) in pastures. Spread at 100kg/ha, the fertiliser boosts batiki grass growth, enabling it to compete more strongly and thus smother (exclude from light) the 2 weed species. Note: Mintweed should be slashed to the ground at the time of fertiliser application. Also the cattle should be excluded from the area for about six weeks to allow the grass to grow.

DAIRY FARMING

Dairy farming is a class of agricultural, or an animal husbandry, enterprise, for long-term production of milk, usually from dairy cows but also from goats, sheep and camels, which may be either processed on-site or transported to a dairy factory for processing and eventual retail sale.

Most dairy farms sell the male calves born by their cows, usually for veal production, or breeding depending on quality of the bull calf, rather than raising non-milk-producing stock. Many dairy farms also grow their own feed, typically including corn, and hay. This is fed directly to the cows, or is stored as silage for use during the winter season.

History

Dairy farming has been part of agriculture for thousands of years. Historically it has been one part of small, diverse farms. In the last century or so larger farms doing only dairy production have emerged. Large scale dairy farming is only viable where either a large amount of milk is required for production of more durable dairy products such as cheese, butter, etc. or there is a substantial market of people with cash to buy milk, but no cows of their own.

Hand milking

Centralized dairy farming as we understand it primarily developed around villages and cities, where residents were unable to have cows of their own due to a lack of grazing land. Near the town, farmers could make some extra money on the side by having additional animals and selling the milk in town. The dairy farmers would fill barrels with milk in the morning and bring it to market on a wagon. Until the late 19th century, the milking of the cow was done by hand. In the United States, several large dairy operations existed in some northeastern states and in the west, that involved as many as several hundred cows, but an individual milker could not be expected to milk more than a dozen cows a day. Smaller operations predominated.

For most herds, milking took place indoors twice a day, in a barn with the cattle tied by the neck with ropes or held in place by stanchions. Feeding could occur simultaneously with milking in the barn, although most dairy cattle were pastured during the day between milkings. Such examples of this method of dairy farming are difficult to locate, but some are preserved as a historic site for a glimpse into the days gone by. One such instance that is open for this is at Point Reyes National Seashore.

Vacuum bucket milking

The first milking machines were an extension of the traditional milking pail. The early milker device fit on top of a regular milk pail and sat on the floor under the cow. Following each cow being milked, the bucket would be dumped into a holding tank. This developed into the Surge hanging milker. Prior to milking a cow, a large wide leather strap called a surcingle was put around the cow, across the cow's lower back. The milker device and collection tank hung underneath the cow from the strap. This innovation allowed the cow to move around naturally during the milking process rather than having to stand perfectly still over a bucket on the floor.

Milking pipeline

The next innovation in automatic milking was the milk pipeline. This uses a permanent milk-return pipe and a second vacuum pipe that encircles the barn or milking parlor above the rows of cows, with quick-seal entry ports above each cow. By eliminating the need for the milk container, the milking device shrank in size and weight to the point where it could hang under the cow, held up only by the sucking force of the milker nipples on the cow's udder. The milk is pulled up into the milk-return pipe by the vacuum system, and then flows by gravity to the milkhouse vacuum-breaker that puts the milk in the storage tank. The pipeline system greatly reduced the physical labor of milking since the farmer no longer needed to carry around huge heavy buckets of milk from each cow.

The pipeline allowed barn length to keep increasing and expanding, but after a point farmers started to milk the cows in large groups, filling the barn with one-half to one-third of the herd, milking the animals, and then emptying and refilling the barn. As herd sizes continued to increase, this evolved into the more efficient milking parlor.

Milking parlors

Innovation in milking focused on mechanizing the milking parlor (known in Australia and New Zealand as a milking shed) to maximize the number of cows per operator which streamlined the milking process to permit cows to be milked as if on an assembly line, and to reduce physical stresses on the farmer by putting the cows on a platform slightly above the person milking the cows to eliminate having to constantly bend over. Many older and smaller farms still have tie-stall or stanchion barns, but worldwide a majority of commercial farms have parlors.

Herringbone and parallel parlors

In herringbone and parallel parlors, the milker generally milks one row at a time. The milker will move a row of cows from the holding yard into the milking parlor, and milk each cow in that row. Once all or most of the milking machines have been removed from the milked row, the milker releases the cows to their feed. A new group of cows is then loaded into the now vacant side and the process repeats until all cows are milked. Depending on the size of the milking parlor, which normally is the bottleneck, these rows of cows can range from four to sixty at a time.

Rotary parlors

In rotary parlors, the cows are loaded one at a time onto the platform as it rotates. The milker stands near the entry to the parlor and puts the cups on the cows as they move past. By the time the platform has completed almost a full rotation, another milker or a machine removes the cups and the cow steps backwards off the platform and then walks to its feed. Rotary cowsheds, as they are called in New

Zealand, started in the 1980s but are expensive compared to Herringbone cowshed - the older New Zealand norm. A rotary is about 25% faster than a herringbone shed for the same number of cows.

Automatic milker take-off

It can be harmful to an animal for it to be over-milked past the point where the udder has stopped releasing milk. Consequently the milking process involves not just applying the milker, but also monitoring the process to determine when the animal has been milked out and the milker should be removed. While parlor operations allowed a farmer to milk many more animals much more quickly, it also increased the number of animals to be monitored simultaneously by the farmer. The automatic take-off system was developed to remove the milker from the cow when the milk flow reaches a preset level, relieving the farmer of the duties of carefully watching over 20 or more animals being milked at the same time. This is a standard procedure in New Zealand.

Fully automated robotic milking

In the 1980s and 1990s, robotic milking systems were developed and introduced (principally in the EU). Thousands of these systems are now in routine operation. In these systems the cow has a high degree of autonomy to choose her time of milking within pre-defined windows. These systems are generally limited to intensively managed systems although research continues to match them to the requirements of grazing cattle and to develop sensors to detect animal health and fertility automatically.Every time the cow enters the stall she is fed and her collar is scanned to record that she was milked.

History of milk preservation methods

Cool temperature has been the main method by which milk freshness has been extended. When windmills and well pumps were invented, one of its first uses on the farm besides providing water for animals was for cooling milk, to extend the storage life before being transported to the town market.

The naturally cold underground water would be continuously pumped into a tub or other containers of milk set in the tub to cool after milking. This method of milk cooling was extremely popular before the arrival of electricity and refrigeration.

Refrigeration

When refrigeration first arrived (the 19th century) the equipment was initially used to cool cans of milk, which were filled by hand milking. These cans were placed into a cooled water bath to remove heat and keep them cool until they were able to be transported to a collection facility. As more automated methods were developed for harvesting milk, hand milking was replaced and, as a result, the milk can was replaced by a bulk milk cooler. 'Ice banks' were the first type of bulk milk cooler.

This was a double wall vessel with evaporator coils and water located between the walls at the bottom and sides of the tank. A small refrigeration compressor was used to remove heat from the evaporator coils. Ice eventually builds up around the coils, until it reaches a thickness of about three inches surrounding each pipe, and the cooling system shuts off. When the milking operation starts, only the milk agitator and the water circulation pump, which flows water across the ice and the steel walls of the tank, are needed to reduce the incoming milk to a temperature below 5 degrees.

This cooling method worked well for smaller dairies, however was fairly inefficient and was unable to meet the increasingly higher cooling demand of larger milking parlors. In the mid-1950s direct expansion refrigeration was first applied directly to the bulk milk cooler. This type of cooling utilizes an evaporator built directly into the inner wall of the storage tank to remove heat from the milk. Direct expansion is able to cool milk at a much faster rate than early ice bank type coolers and is still the primary method for bulk tank cooling today on small to medium sized operations.

Another device which has contributed significantly to milk quality is the plate heat exchanger (PHE). This device utilizes a number of specially designed stainless steel plates with small spaces between them. Milk is passed between every other set of plates with water being passed between the balance of the plates to remove heat from the milk. This method of cooling can remove large amounts of heat from the milk in a very short time, thus drastically slowing bacteria growth and thereby improving milk quality. Ground water is the most common source of cooling medium for this device. Dairy cows consume approximately 3 gallons of water for every gallon of milk production and prefer to drink slightly warm water as opposed to cold ground water. For this reason, PHE's can result in drastically improved milk quality, reduced operating costs for the dairymen by reducing the refrigeration load on his bulk milk cooler, and increased milk production by supplying the cows with a source of fresh warm water.

Plate heat exchangers have also evolved as a result of the increase of dairy farm herd sizes in the United States. As a dairyman increases the size of his herd, he must also increase the capacity of his milking parlor in order to harvest the additional milk. This increase in parlor sizes has resulted in tremendous increases in milk throughput and cooling demand. Today's larger farms produce milk at a rate which direct expansion refrigeration systems on bulk milk coolers cannot cool in a timely manner. PHE's are typically utilized in this instance to rapidly cool the milk to the desired temperature (or close to it) before it reaches the bulk milk tank. Typically, ground water is still utilized to provide some initial cooling to bring the milk to between 55 and 70 °F (21 °C). A second (and sometimes third) section of the PHE is added to remove the remaining heat with a mixture of chilled pure water and propylene glycol. These chiller systems can be made to incorporate large evaporator surface areas and high chilled water flow rates to cool high flow rates of milk.

Milking operation

Milking machines are held in place automatically by a vacuum system that draws the ambient air pressure down from 15 to 21 pounds per square inch (100 to 140 kPa) of vacuum. The vacuum is also used to lift milk vertically through small diameter hoses, into the receiving can. A milk lift pump draws the milk from the receiving can through large diameter stainless steel piping, through the plate cooler, then into a refrigerated bulk tank.

Milk is extracted from the cow's udder by flexible rubber sheaths known as liners or inflations that are surrounded by a rigid air chamber. A pulsating flow of ambient air and vacuum is applied to the inflation's air chamber during the milking process. When ambient air is allowed to enter the chamber, the vacuum inside the inflation causes the inflation to collapse around the cow's teat, squeezing the milk out of teat in a similar fashion as a baby calf's mouth massaging the teat. When the vacuum is reapplied in the chamber the flexible rubber inflation relaxes and opens up, preparing for the next squeezing cycle.

It takes the average cow three to five minutes to give her milk. Some cows are faster or slower. Slow-milking cows may take up to fifteen minutes to let down all their milk. Milking speed is only minorly related to the quantity of milk the cow produces - milking speed is a separate factor from milk quantity; milk quantity is not determinative of milking speed. Because most milkers milk cattle in groups, the milker can only process a group of cows at the speed of the slowest-milking cow. For this reason, many farmers will group slow-milking cows so as not to stress the faster milking cows.

The extracted milk passes through a strainer and plate heat exchangers before entering the tank, where it can be stored safely for a few days at approximately 42 °F (6 °C). At pre-arranged times, a milk truck arrives and pumps the milk from the tank for transport to a dairy factory where it will be pasteurized and processed into many products.

Management of the herd

Modern dairy farmers use milking machines and sophisticated plumbing systems to harvest and store the milk from the cows, which are usually milked two or three times daily. In New Zealand some farmers seeking a better life style, are milking only once per day, trading a slight reduction in production of milk for increased leisure time. During the summer months, cows may be turned out to graze in pastures, both day and night, and are brought into the barn to be milked.

Barns may also incorporate tunnel ventilation into the architecture of the barn structure. This ventilation system is highly efficient and involves opening both ends of the structure allowing cool air to blow through the building. Farmers with this type of structure keep cows inside during the summer months to prevent sunburn and damage to udders. During the winter months the cows may be kept in the barn, which is warmed by their collective body heat. Even in winter, the heat produced by the cattle requires the barns to be ventilated for cooling purposes. Many modern facilities, and particularly those in tropical areas, keep all animals inside at all times to facilitate herd management.

Housing can be either loose in stalls (called cow cubicles in UK). There is little research available on dimensions required for cow stalls, and much housing can be out of date, however increasingly companies are making farmers aware of the benefits, in terms of animal welfare, health and milk production.

Farmers typically grow their own food for their cattle. This may include corn, and "hay", which includes alfalfa, timothy, and clover.

In the southern hemisphere such as in Australia and New Zealand, cows spend most of their lives outside on pasture, although they may receive supplementation during periods of low pasture availability. Typical supplementary feeds in Australasia are hay, silage or ground maize. The trend in New Zealand is towards feeding cows on a concrete pad to prevent loss of feed by trampling. In New Zealand slower growing winter pasture is rationed. It is carefully controlled by light weight portable electric break feeding fences run on mains power that can be easily repositioned.

Concerns

Animal waste from large dairies

As measured in phosphorus, the waste output of 5,000 cows roughly equals a municipality of 70,000 people. In the U.S., dairy operations with more than 1,000 cows meet the EPA definition of a CAFO (Concentrated Animal Feeding Operation), and are subject to EPA regulations. For example, in the San Joaquin Valley of California a number of dairies have been established on a very large scale. Each dairy consists of several modern milking parlor set-ups operated as a single enterprise. Each milking parlor is surrounded by a set of 3 or 4 loafing barns housing 1,500 or 2,000 cattle. Some of the larger dairies have planned 10 or more series of loafing barns and milking parlors in this arrangement, so that the total operation may include as many as 15,000 or 20,000 cows. The milking process for these dairies is similar to a smaller dairy with a single milking parlor but repeated several times. The size and concentration of cattle creates major environmental issues associated with manure handling and disposal, which requires substantial areas of cropland (a ratio of 5 or 6 cows to the acre, or several thousand acres for dairies of this size) for manure spreading and dispersion, or several-acre methane digesters. Air pollution from methane gas associated with manure management also is a major concern. As a result, proposals to develop dairies of this size can be controversial and provoke substantial opposition from environmentalists including the Sierra Club and local activists.

The potential impact of large dairies was demonstrated when a massive manure spill occurred on a 5,000-cow dairy in Upstate New York, contaminating a 20-mile (32 km) stretch of the Black River, and killing 375,000 fish. On Aug. 10, 2005, a manure storage lagoon collapsed releasing 3,000,000 US gallons (11,000,000 l; 2,500,000 imp gal) of manure into the Black River. Subsequently the New York Department of Environmental Conservation mandated a settlement package of $2.2 million against the dairy.

When properly managed, dairy and other livestock waste, due to its nutrient content (N, P, K), makes an excellent fertilizer promoting crop growth, increasing soil organic matter, and improving overall soil fertility and tilth characteristics. Most dairy farms in the United States are required to develop nutrient management plans for their farms, to help balance the flow of nutrients and reduce the risks of environmental pollution. These plans encourage producers to monitor all nutrients coming onto the farm as feed, forage, animals, fertilizer, etc. and all nutrients exiting the farm as product, crop, animals, manure, etc. For example, a precision approach to animal feeding results in less overfeeding of nutrients and a subsequent decrease in environmental excretion of nutrients, such as phosphorus. In recent years, nutritionists have realized that requirements for phosphorus are much lower than previously thought. These changes have allowed dairy producers to reduce the amount of phosphorus being fed to their cows with a reduction in environmental pollution.

In New Zealand the average dairy farmer has 500 cows which are milked twice per day taking about 2 hours per milking. All cows are grazed on grass pasture. Dung and urine from the milking shed is flushed into drains by large high pressure hoses, which lead to open pits. Over time the solid matter settles to the bottom and is removed by truck about once per year. The relatively clean water is allowed to percolate through natural swamp and creeks back to major rivers. Local authorities check that water entering rivers meets minimum standards. Farmers are fined and forced to change their systems to meet the standards. Repeated infringement means the farm is closed down. Local councils often supply large numbers of native swamp plants grown in their own nurseries to farmer at a low cost. Planting is also done by environmental groups, and schools as part of their science program and groups of unemployed on relief work.

Use of hormones

It is possible to maintain higher milk production by injecting cows with growth hormones known as recombinant BST or rBST, but this is controversial due to its effects on animal and possibly human health. The European Union, Japan, Australia, New Zealand and Canada have banned its use due to these concerns.

In the US however, no such prohibition exists, and approximately 17.2% of dairy cows are treated in this way. The U.S. Food and Drug Administration states that no "significant difference" has been found between milk from treated and non-treated cows but based on consumer concerns several milk purchasers and resellers have elected not to purchase milk produced with rBST.

Animal welfare

The practice of dairy production in a factory farm environment has been criticized by animal welfare activists. Some of the ethical complaints regarding dairy production cited include how often the dairy cattle must remain pregnant, the separation of calves from their mothers, how dairy cattle are housed and environmental concerns regarding dairy production.

The production of milk requires that the cow be in lactation, which is a result of the cow having given birth to a calf. The cycle of insemination, pregnancy, parturition, and lactation, followed by a "dry" period of about two months of forty-five to fifty days, before calving which allows udder tissue to regenerate. A dry period that falls outside this time frames can result in decreased milk production in subsequent lactation. Dairy operations therefore include both the production of milk and the production of calves. Bull calves are either castrated and raised as steers for beef production or veal.

An important part of the dairy industry is the removal of the calves off the mother's milk after the three days of needed colostrum, allowing for the collection of the milk produced. In order for this to take place, the calves are fed milk replacer, a substitute for the whole milk produced by the cow. Milk replacer is generally a powder, which comes in large bags, and is added to precise amounts of water, and then fed to the calf via bucket or bottle.

Milk replacers are classified by three categories: protein source, protein/fat (energy) levels, and medication or additives (e.g. vitamins and minerals). Proteins for the milk replacer come from different sources; the more favorable and more expensive all milk protein (e.g. whey protein- a bi product of the cheese industry) and alternative proteins including soy, animal plasma and wheat gluten. The ideal levels for fat and protein in milk replacer are 10-28% and 18-30%, respectively. The higher the energy levels (fat and protein), the less starter feed (feed which is given to young animals) the animal will consume. Weaning can take place when a calf is consuming at least two pounds of starter feed a day and has been on starter for at least three weeks. Milk replacer has climbed in cost US$15-20 a bag in recent years, so early weaning is economically crucial to effective calf management.

Because of the danger of infection to humans, it is important to maintain the health of milk-producing cattle. Common ailments affecting dairy cows include infectious disease (e.g. mastitis, endometritis and digital dermatitis), metabolic disease (e.g. milk fever and ketosis) and injuries caused by their environment (e.g. hoof and hock lesions).

Lameness is commonly considered one of the most significant animal welfare issues for dairy cattle, and is best defined as any abnormality that causes an animal to change its gait. It can be caused by a number of sources, including infections of the hoof tissue (e.g. fungal infections that cause dermatitis) and physical damage causing bruising or lesions (e.g. ulcers or hemorrhage of the hoof). Housing and management features common in modern dairy farms (such as concrete barn floors, limited access to pasture and suboptimal bed-stall design) have been identified as contributing risk factors to infections and injuries.

Market

Worldwide

There is a great deal of variation in the pattern of dairy production worldwide. Many countries which are large producers consume most of this internally, while others (in particular New Zealand), export a large percentage of their production. Internal consumption is often in the form of liquid milk, while the bulk of international trade is in processed dairy products such as milk powder.

Most milk-consuming countries have a local dairy farming industry, and most producing countries maintain significant tariffs to protect domestic producers from foreign competition but, the largest dairy exporting country, does not apply any subsidies to dairy production.

The milking of cows was traditionally a labor-intensive operation and still is in less developed countries. Small farms need several people to milk and care for only a few dozen cows, though for many farms these employees have traditionally been the children of the farm family, giving rise to the term "family farm".

Advances in technology have mostly led to the radical redefinition of "family farms" in industrialized countries such as Australia, New Zealand, and the United States. With farms of hundreds of cows producing large volumes of milk, the larger and more efficient dairy farms are more able to weather severe changes in milk price and operate profitably, while "traditional" very small farms generally do not have the equity or cash flow to do so. The common public perception of large corporate farms supplanting smaller ones is generally a misconception, as many small family farms expand to take advantage of economies of scale, and incorporate the business to limit the legal liabilities of the owners and simplify such things as tax management.

Before large scale mechanization arrived in the 1950s, keeping a dozen milk cows for the sale of milk was profitable. Now most dairies must have more than one hundred cows being milked at a time in order to be profitable, with other cows and heifers waiting to be "freshened" to join the milking herd . In New Zealand the average herd size, for the 2009/2010 season, is 376 cows.

Worldwide, the largest milk producer is the European Union with its present 27 member countries, with more than 153,000,000 metric tons (151,000,000 long tons; 169,000,000 short tons) in 2009 (more than 95% cow milk). By country, the largest producer is India (more than 55% buffalo milk), the largest cow milk exporter is New Zealand, and the largest importer is China.

European Union

The European Union with its present 27 member countries is the largest milk producer in the world. The largest producers within the EU are Germany and France.

Dairy production in the EU is heavily distorted due to the Common Agricultural Policy - being subsidized in some areas, and subject to production quotas in other.

Israel

The dairy farm on Sa'ad was the Israeli leader in 2011 for productivity with an average of 13,785 litres (3,032 imp gal; 3,642 US gal) per head that year. A dairy cow named Kharta, was the world record holder giving 18,208 litres (4,005 imp gal; 4,810 US gal) liters of milk. The 954 Israeli dairy farms achieved a world leading average production of 11,775 litres (2,590 imp gal; 3,111 US gal) a year per head, while the national average per head was 10,336 litres (2,274 imp gal; 2,730 US gal). Israeli consumption is lower than other western countries with an average of 180 litres (40 imp gal; 48 US gal) per person.

United States

In the United States, the top six dairy states are, in order by total milk production; California, Wisconsin, Idaho, New York, Michigan, and Pennsylvania. Dairy farming is also an important industry in Florida, Minnesota, Ohio and Vermont. There are 65,000 dairy farms in the United States.

Pennsylvania has 8,500 farms with 555,000 dairy cows. Milk produced in Pennsylvania yields an annual revenue of about US$1.5 billion.

Milk prices collapsed in 2009. Senator Bernie Sanders accused Dean Foods of controlling 40% of the country's milk market. He has requested the United States Department of Justice to pursue an anti-trust investigation. Dean Foods says it buys 15% of the country's raw milk. In 2011, a federal judge approved a settlement of $30 million to 9,000 farmers in the Northeast.

Herd size in the US varies between 1,200 on the West Coast and Southwest, where large farms are commonplace, to roughly 50 in the Midwest and Northeast, where land-base is a significant limiting factor to herd size. The average herd size in the U.S. is about one hundred cows per farm.

2

Housing of Cattles

Efficient management of cattle is incomplete without a well planned and adequate housing. Considering the economy, productiveness and protection from inclement weather necessitate the confinement of cattle in houses, sheds or barn. Improper planning in the arrangement of cattle housing may result in additional labour charges and thus curtail the profit of the owner. In the past 10 years or so there has been a revolution in keeping farm livestock. Intensification of cattle farming has taken place and large number of cattle are confined to small lots.

In India, sufficient attention has yet not been paid to this important aspect of cattle management. We still find different types of cattle houses constructed without careful designing and planning. During erection of a cattle house, care should be taken to provide comfortable accommodation for an individual cattle. No less attention should be paid to the durability, proper sanitation and arrangement for the production of clean milk under convenient and economic conditions. Economy in their initial cost and maintenance of the building should also receive special attention so as to reduce manual labour requirement in the efficient management.

Location of Dairy Buildings

The points which should be considered before the erection of dairy buildings are as follows:

1. Topography and drainage

A dairy building should be located at an higher elevation than the surrounding ground to offer good slope for rainfall and drainage for the wastes of dairy to avoid the stagnation within it. A level led area requires less site preparation and thus lesser cost of building. Low lands and depressions and proximity to places of bad odour should be avoided.

2. Soil type

Fertile soil should be spared for cultivation. Foundation soils as far as possible should not be too dehydrated or desiccated. Such a soil is susceptible to considerable swelling during rainy season and exhibit numerous cracks and fissures.

3. Exposure to sun and protection from wind

A dairy building should be located to maximum exposure to the sun in the north and minimum exposure to the sun in the south and have protection from prevailing strong, hot or cold wind currents. Buildings should be so constructed that direct sunlight can reach the platforms, gutters and mangers in the cattle shed.

As far as possible, long axis of the dairy barns should beset in the North -South direction to have the maximum benefit of the sun.

4. Accessibility

Easy accessibility to the buildings is always desirable. Situation of a cattle shed by the side of the main road preferably at a distance of about 100 meters should be aimed at.

5. Durability and attractiveness

It is always attractive when the building open up to a scenic view and to the grandeur of the scenery. Along with this, durability of the structure is obviously an important criteria in buildingn a dairy.

6. Water supply

Abundant supply of fresh, clean and soft water should .be available at a cheap rate.

7. Surroundings

Areas infested with wild animals or dacoit should be avoided. Narrow gates, higher manger curbs, loose hinges, protruding nails, smooth finished floor in the areas where the cow moves and other such hazards should be eliminated.

8. Marketing

Dairy buildings should only be located in those areas from where the owner can sell his products profitably and regularly. He should be in a position to satisfy the needs of the farm within no time and at a reasonable price.

9. Labour

Honest, economic and regular supply of labour is available.

10. Electricity

Electricity is the most important sanitary method of lighting a dairy. Since a modem dairy always handles electric equipments which are also economical, it is desirable to have an adequate supply of electricity.

11. Facilities, labour, and feed

Cattle yards should be so constructed and situated in relation to feed storages, hay stacks, silo and manure pits as to effect the most efficient utilization of labour. Sufficient space per cow and well arranged feeding mangers and resting areas contribute not only to greater milk yield of cows and make the work of operator easier but also minimizes the food expenses. The relative position of food stores should be quite adjacent to the cattle barn. Important features of the feed stores are given below.

- Feed storages should be located near the center of the cow barn.
- Milk house should be located almost at the centre of the barn.
- Centre cross alley should be well designed with reference to feed storages, the stall area and the milk house.

Design Consideration for Cattle Houses

When designing cattle houses the following points must be considered.

1. Shelter

The basic justification of livestock shelter is that it should alter or modify the environment for the benefit of animals enclosed in it. The animal shelter should normally buffer the extremes of climate to reduce the peak stress on the animals housed therein. The main climatic factors from which protection is required are high and low ambient temperatures, environmental humidity, heat, wind and rain.

2.Animal production

Animal growth, production and reproduction are not generally influenced by slight and normal seasonal variations in environmental factors. But when animals are exposed to extreme conditions for longer periods, the performance of animals is impaired. This may also result in death in extreme cases.

Thus, animal houses should be so designed that they are cheaper and can afford protection from extreme weather conditions. Expensive fittings and design on a permanent basis should not be resorted to face only brief periods of climatic extremes.

TYPES OF HOUSING

The most widely prevalent practice in this country is to tie the cows with rope on a Katchafloor except some organized dairy farms belonging to government, co-operatives or military where proper housing facilities exist. It is quit easy to understand that unless cattle are provide with good housing facilities, the animals will move too far in or out of the standing space, defecating all round and even causing trampling and wasting of feed by stepping into the mangers. The animals will be exposed to extreme weather conditions all leading to bad health and lower production.

Dairy cattle may be successfully housed under a wide variety of conditions, ranging from close confinement to little restrictions except at milking time. However, two types of dairy barns are in general use at the present time.

1. The loose housing barn in combination with some type of milking barn or parlor.
2. The conventional dairy barn.

Loose housing system

Loose housing may be defined as a system where animals are kept loose except milking and at the time of treatment. The system is most economical.

Advantages

1. There is flexibility of use of buildings. The bam can be utilized for various animal husbandry enterprises with light modifications.
2. The manure is well preserved. All droppings and urine, with generous amount of bedding, are kept under a roof and well protected, thus reducing losses.
3. It is easy to detect cows in heat, because the cows are free to move most of the time.
4. The building construction can be of an inexpensive type.
5. Injuries to cows are kept at a minimum. There are no gutters to injure udders and legs and no stanchion curbs to injure knees. Cow lie on a deeply bedded surface.
6. Milking is done in a clean, compact and efficient milking room. Cows rnay be thoroughly washed and cleaned occasionally and before coming to milking room their udders are washed with clean water.
7. There is flexibility of herd size. Room for more cows may easily be made by extending the hay and silage banks and by enlarging the pens.

Disadvantages

1. The herd cannot be displayed easily.
2. More bedding material is needed.
3. Cows in heat disturb the herd.
4. A larger building is needed. The total floor area for a loose housing barn will be 10 -25% more than for a stanchion barn.

The floor and manger space requirement under loose housing system for different types of cattle are given in the following table 1.

The animal sheds should have proper facilities for milking barns, calf pens, calving pens and arrangement for store room etc. In each shed there should be arrangement for feeding, drinking area and loafing area.

The shed may be cemented or brick paved, but in any case it should be easy to clean. The floor should be rough, so that the animals do not slip. The drains in the shed should be shallow and preferably covered with removable tiles. The drain should have a gradient of 1" (inch) for every 10'(feet) length. The roof may be of corrugated cement sheet, asbestos. or brick and rafters. The cement concrete roofing are too expensive. Inside the open unpaved area, it is always desirable to plant some good shady trees for protection against direct cold wind in winter and to keep cool in summer.

Cattle Shed

The entire shed should be surrounded by a boundary wall of. 5" height from three side and manger etc., on one side. The feeding area should be provided with 2 to 2 ½ feet of manger space per cow. All along the manger, there shall be 10" wide water trough to provide clean, even, available drinking water. The water trough thus constructed will also minimize the loss of fodders during feeding. Near the manger, under the roofed house 5' wide floor should be paved with bricks having a little slope.

Beyond that, there should be open unpaved area (40'X35') surrounded by 5' wall with one gate. It is preferable that animals face north when they are eating fodder under the shade. During cold wind in winter the animals will automatically lie down to have the protection from the walls.

Fig: Cattle Shed

Shed for calves

On one side of the main cattle shed there be fully covered shed "10'x15' to accommodate young calves. Such sheds with suitable partitioning, may also serve as calving pen under adverse climatic conditions. Beyond this covered area there should be a 20'x10' open area having boundary wall so that calves may move there freely.

In this way both calf and cattle sheds will need in all 50'x50' area for 20 adult cows and followers. If one has limited resources, he can build ordinary, Katcha/ semikatcha boundary walls but feeding and water trough should be cemented ones.

Loafing area

Loafing area is also known as lounging area. It is built along lines similar to the ordinary implement shed, using a rather high front entrance on the site next to the feeding barn. Allow 50- 75 sq. feet space per cow, exclusive of the space set aside for feeding. There should be few or no post, to make cleaning easy. The floor can be of earth. In cold places, the cleaning is done once or twice a year, but in India, the cleaning should be done more frequently, -especially when there is more rainfall. The cows will use loafing area more freely if it adequately lighted.

Milking structures

The milking room may be constructed according to the size of the herd and the amount of capital available. A herd of 20-30 cows can be milked in 10 cows milking barn by making 2-3 changes of cows at each milking. Generally grains are fed during milking and, therefore, this type of milking barn will allow enough time for eating grains.

In the parlour type of milking structures where raised platform are used for milking, grain eating time may be limited. Locate the milking room in such a place so that the doors will open into both the loafing and feeding areas.

Special areas

Maternity stalls may be formed by gate partitions in the loafing areas. Permanent stalls may be provided if desired. These pens should be about 10'x 10' in size. Calf pens may be located near the milking room and milk house to facilitate feeding and watering.

Loose housing is a structure for dairy cows when cows run loose in a pen or an open area except when they are being milked. A well bedded area is provided for them to lie down. Therefore, this type of housing can be built with considerably less expenditure and, moreover, cows are in their natural state of comfort. Undoubtedly, this type of housing is better than the stanchion type where the cows are tied at all times except when they are on grazing. One care that is to be taken with this type of housing is to dehorn all the cows and to remove the extremely bossy cows which disturbs the others while eating.

Conventional Dairy Barn

The conventional dairy barns are comparatively costly and are now becoming less popular day by day. However, by this system cattle are more protected from adverse climatic condition.

The following barns are generally needed for proper housing of different classes.

- Dairy stock in the farm

Cow Sheds

Cow sheds can be arranged in a single row if the numbers of cows are small. Say less than 10 or in a double row if the herd is a large one. Ordinarily, not more than 80 to 100 cows should be placed in one building. In double row housing, the stable should be so arranged that the cows face out (tails to tail system) or face in (head to head system) as preferred.

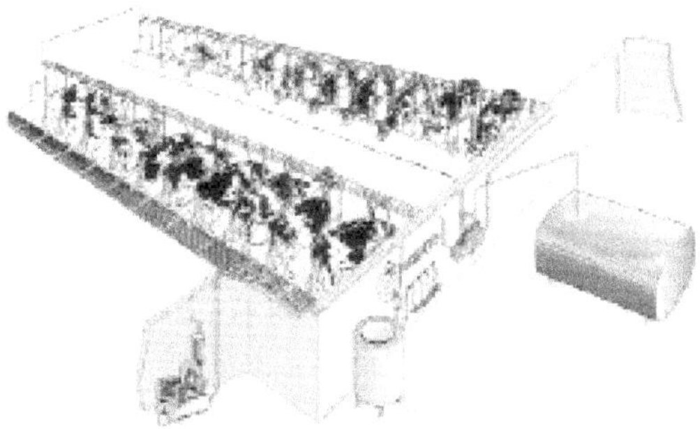

Fig : Cow sheds

The advantages of tail to tail system are mentioned below.

1. Under the average conditions, 125-150 man hours of labour are required per cow per year.
2. In cleaning and milking of the cows the wide middle alley is of great help.
3. Lesser danger of spread of diseases from animal to animal.
4. Cows can always get fresh air from outside.
5. The head milkman can inspect a greater number of other milk man while milking. This is possible because milkmen will be milking on both sides of the head milkman or *gowala.*
6. Any sort of minor disease or any change in the hind quarters of the animals can be detected easily.

The advantages of lace to face system are also cited here.

1. Cows make a better showing for visitors when heads are together.
2. The cows feel easier to get into the stall.
3. Sun rays shine in the gutter where they are needed most.
4. Feeding of cows is easier, both rows can be fed without back tracking.
5. This system is better for narrow barns.

Floor

The inside floor of the barn should be of some impervious material which can be easily kept clean and dry and is not slippery.

Paving with bricks and grooved cement concrete floor which is still better and can also serve ones purpose. The surface of the cow shed should be laid with a gradient of 1” (inch) to 1.5” from manger to gutter. An overall floor space of 65 to 70 sq ft per adult cow should be satisfactory.

Walls

The inside of the walls should have a smooth, hard finish of cement which will not allow any lodgement of dust and moisture. Comers of the walls should be round. For plains, low walls about 4 -5 feet in height and roofs supported by masonary work or iron pillars will be more suitable. The open space in between supporting pillars will serve for light and air circulation.

Roof

Roofs of the barns may be made of asbestos sheet or tiles. Corrugated iron sheets have the disadvantage of creating extreme fluctuation in the inside temperature of the barn in different seasons. However, iron sheet with aluminium painted tops to reflect sunrays and bottoms provided with wooded insulated ceilings can also achieve the objective. A height of 8 feet at the sides and 15 feet at the ridge will be sufficient to give the necessary air space to the cows. An adult cow requires at least about 800 cubic feet of air space under tropical conditions. To make ventilation more effective, a continuous ridge ventilation should be provided.

Stall design

The two main types of dairy barn stalls which are commonly adopted are described below.

Stanchion stall

It is one of the standard dairy cow stalls. It is equipped with a stanchion for fastening the cow in place. Usually there is a stall partition in the form of a curved pipe between the stalls to keep the cows in place and to protect their udders and teats from being stepped on by other cows.

The stanchion should be so constructed and arranged in order to allow the cows the greatest possible freedom. There should be several links of chain at the top and bottom of the stanchions and sufficient room on each side of it to permit the animal to move its head from side to side. It is important to provide for the comfort of the cows and to line them up in such a way so that most of the droppings and urine go to the gutter. Practically, it is not possible to fit every cow to her stall properly. To compensate this, many stanchions have adjustments so that they can be set forward if the cow is too large for the stall or backwards if the cow is too small. The cow can be fastened easily and quickly with the stanchions and is held more closely in place than other types of ties. However, she is held more rigidly and, therefore, the stanchions is less comfortable than other types of fasteners.

The advantages of stanchion barn are given here.

- Individual feeding is possible.
- The herd can be easily displayed.
- Less bedding material is required.
- Cows in heat do not disturb others.
- Well suited for two storey type of structure permitting storage of hay, bedding and grain conveniently located overhead.

The main disadvantages of this system are cited here.

- Regular cleaning requires more man hours.
- Less chance to increase the herd size of barn.
- Barn can be suitable only for dairy farming without major re-modelling.

- Separate manure pit and liquid manure tank are required to preserve the manure properly
- It is more difficult to detect cows in heat.
- Building construction is generally more expensive.
- Cows are more likely to be injured.
- The cows are often not comfortable as they are in the loose " housing barn.

Tie stall

The tie stall requires a few inches longer and wider space than the stanchion stall. It is designed to provide greater comfort to the cow. In addition to larger size, the chain tie gives the cow more freedom. Instead of the stanchion, there are two arches, one on each side of the neck of the cow. The cow is fastened by means of the rings fitted loosely on the arch pipes and connected to a chain which snaps to the neck strap on the cow. The correct space between the arches is 10 to 12 inches. This prevents the cow from moving too far forward in the stall.

It is important that in this type of stall, the arches and all other stall part; are kept lower than the height of the cows. The cow has more freedom in the tie stall than in the stanchion. Large cows and those with large udders get along better in them because of the freedom they enjoy. It is not desirable to have a tie chain in small stalls.

Manger

Cement concrete continuous manger with removable partitions is the best from the point of view of durability and cleanliness. A height of 1 '-4" for a high front manger and 6" to 9" for a low front manger is considered sufficient. Low front mangers are more comfortable for cattle but high front. mangers prevent feed wastage. The height at the back of the manger should be kept at 2'-6" to 3". An overall width of 2' to 2 1/2' is sufficient for a good manger.

Alleys

The central walk should have a width of 5'-6' exclusive of gutters when cows face out, and 4'-5' when they face in. The feed alley, in case of a face out system should be 4' wide, and the central walk should show a slope of 1" from the center towards the two gutters running parallel to each other, thus forming a crown at the center.

Manure gutter

The manure gutter should be wide enough to hold all dung without getting blocked, and be easy to clean/ Suitable dimensions are 2" width with a cross-fall of !" away from standing. The gutter should have a gradient of 1" for every 10' length. This will permit a free flow of liquid excreta.

Doors

The doors of a single range cowshed should be 5" wide with a height of 7', and for double row shed the width should not be less than 8" to 9'. All doors of the barn should lie flat against the external wall when fully open.

Calving Boxes

Allowing cows to calve in the milking cowshed is highly undesirable and objectionable. It leads to in sanitary in milk production and spread of disease like contagious abortion in the herd. Special accommodation in the form of loose-boxes enclosed from all sides with a door should be furnished to all parturient cows. It should have an area of about 100 to 150 sq.ft. With ample soft bedding, it should be provided with sufficient ventilation through windows and ridge vent.

Isolation Boxes

Animals suffering from infectious disease must be segregated soon from the rest of the herd. Loose boxes of about 150 sq.ft are very suitable for this purpose. They should be situated at some distance from the other barns. Every isolation box should be self contained and should have separate connection to the drainage disposal system.

Sheds for young stocks

Calves should never be accommodated within the cow shed. The calf house must have provision for day light ventilation and proper drainage. Damp and ill-drained floors cause respiratory trouble in calves to which they are susceptible. For an efficient management and housing, the young stock should be divided into three groups, viz young calves aged up to one year, bull calves i.e. the male calves over one year and the heifers over one year. Each group should be sheltered in a separate house or shed. For the young calves, the shed should be located quite close to the cowshed. Each calf shed should have an open paddock or exercise yard. An area of 100 square feet per head of a stock of 10 calves and an increase of 50 square feet for every additional calf will make a good paddock.

In the indoor rearing system, the calf house should be provided adequate shelter and be easily cleaned. The house should keep the young calves sufficiently cool in summer. This can be accomplished by constructing a simple roof over a concrete base or a well drained site. The roof should be at least 9 feet high at the canes. Side walls are not necessary at the humid tropics as long as the roof has wide overhand to keep out driving rains. Under the roof the provision should be made for individual feeding, separate pens, etc. Drinking water should be provided in the pen itself. The floor should be dry and impervious to moisture and a light bedding of straw is preferred.

Heifers reared in yards should also have access to clean and cool water, shade, minerals and forage. The shed for heifers (female calves over 12 months age) should be oriented East-West and approximately 10 feet in height. The shed should be constructed in a single row of sheds facing each other.

BULL OR BULLOCK SHED

Safety and ease in handling a comfortable shed protection from weather and a provision for exercise are the key points while planning accommodation for bulls or bullocks. A bull should never be kept in confinement particularly on hard floors. Such a confinement without adequate exercise leads to overgrowth of the hoofs creating difficulty in mounting and loss in the breeding power of the bull. A loose box with rough cement concrete floor about 15' by 10' in dimensions having an adequate arrangement of light and ventilation and an entrance 4' in width and 7' in height will make a comfortable housing for a bull. The shed should have a manger and a water trough.

If possible, the arrangement should be such that water and feed can be served without actually entering the bull house. The bull should have a free access to an exercise yard provided with a strong fence or a boundary wall of about 2' in height, i.e., too high for the bull to jump over. From the bull yard, the bull should be able to view the other animals of the herd so that it does not feel isolated. The exercise yard should also communicate with a service crate via a swing gate which saves the use of an attendant to bring the bull to the service crate.

Ancillary Structure on Dairy Farm

Some other important structures which are very useful in a cattle dairy are described below.

Stores

Stores for concentrate feeds, hay, straw, equipments, etc. are essential on animal farms. The size and type of buildings for stores can be decided on the basis of the quantity of feeds and fodders that are to be stored at a time on the farm. There should be one main concentrate store-cum-feed mixing room at a distant place, but it is advisable to have small ration room nearer to the milking barn for storing prepared concentrate mixture temporarily for a day or two. On smaller farms perhaps one feed store near the milking barn will be sufficient. The feed room should be made damp and rodent proof.

Silos

Silage is an important item of feed for dairy animals. Silage is obtained by ensiling chaffed green fodders in silos. Tower, pit and trench silos are in use, but under Indian conditions, trench silos will be more practicable and convenient. The silos should be made nearer to cow shed as far as possible.

A movable sturdy wooden or metal fence or any other self-feeding device should be fixed to the open end of the trench soils so that the cows can feed on silage at will, thus saving the time and labour otherwise required for transportation and feeding of silage in the cowsheds. The trench for the silo can be a simple excavation made in a hard ground and then the floor and sides hardened by ramming and ground plastering. Cement concrete or brick or stone slab lined trenches, however, are longer lasting and seepage proof.

Hay or straw sheds

The common dry fodders such as hay or crop by-products are available only at harvesting season, so one has to store these items for use throughout the year. Space requirement shall depend upon the manner in which the dry fodder is stored. Adult stock consume about 6 kg of dry fodder per day while young stock consume about 3 kg per day. Annual quantity of hay needed can be worked out on the basis of number of days in a year when hay is required. The size of the shed needed can be thus calculated.

Shed preferably with walls on three sides are better for storing straw or hay. Storing hay or straw in the open results in excessive wastage in the form of spoilage and deterioration. Sometimes, a simple framed shed with gabled roof is used. The hay shed should be situated away from the animal shed because of fire hazards. On larger farms the hay shed can be converted into self-feeding hay banks by placing movable wooden partition on one side edge of the hay shed. The animals can eat hay at will which can be accessible through this wooden partitions and thus do not spoil the hay excessively.

Milk house

Milk house is an important part of a dairy building. This is the place where milk is collected, weighed and stored in cans for brief periods before delivering to the milk plants. Dimensions of a milk house depends upon the quantity of milk handled daily. Based on the milk quantity to be handled every day, the floor space requirement is given in the table -2.

Table 2: Floor space requirement for milk house

Sr. No	Daily milk	Floor space	Remarks
	output (litres)	(metre)	
1.	Below 100	3.7 × 3.0	Additional 0.37
2.	100-200	3.7 × 3.7	Sq.m for every 40
3.	200-450	3.7 × 4.4	Litres output over
4.	450-700	3.7 × 5.1	And above 700 litres

The floor space can accommodate the milk recording equipment, milk cooling device in the form of bulk cooler, can racks, milking pail rack, sinks, washing outfits and furniture for the milk recorder. On very large farms the different components of the milk house i.e. recording cum-milk cooling room, milk utensils and equipments room and washing room may be constructed as separate units.

The doors and the window of the milk house should be made fly-proof. The flooring of the room should be of cement concrete, impervious and reinforced with iron strips located at suitable distances to make it hard wearing. The walls should be lined with white glazed tiles up to a height of 1.5 m. A suitable platform or concrete slab (like that of a laboratory bench) should be provided for keeping the milk testing apparatus.

Handling yards

On larger farms, it is advantageous and economical to construct one flexible handling yard that can be used for the three processes of animal handling, namely collecting, filling and control of individual animal for weighing, vaccination and marketing. The size and number of different component pens depend on the herd of the stock handled. On small farms, provisions of simple trevish cattle crush is sufficient for controlling the animals.

Manure pits

A lot of dung, urine and bedding is found in cattle shed. If this is properly collected and converted into manure, it will prove very valuable to the farm fields.

About 4 or 5 manure pits of 4 x 2 x 1 metre dimensions may be dug out at a place nearer to the fields but far away (at least 12 metres away) from cattle houses. Dung, leftover fodders and other farm wastes are to be taken away from the animal houses in a cart and dumped into the manure pit.

IMPROVING THE EXISTING TRADITIONAL CATTLE SHED

Farmers in India keep their animals under greatly diverse condition such as tethering under a tree, tethering them very close to their living quarters, enclosing them within partially sheltered wood -fenced paddock, or even making small sheds especially for cattle near the living quarters. Under such system of housing, the animals have to face two problems namely overcrowding and remaining standing on a mixture of earth, excreta, urine and water. As a result animals cannot lie down comfortably.

When animals are dirty, not only their wounds (especially those on limbs) heal slowly but the animals may also fall sick easily. And sick animals mean loss of production and, thus loss of money to the farmers. Calves are also affected by the unhygienic surroundings. They may be attacked by several kinds of parasites and vermin's. These conditions often result in heavy calf mortality and there is loss of money again to the farmer.

It is difficult to obtain good manure from the dirty cattle sheds as the dung gets mixed up with earth and urine. The farmer neither can use such a material as manure on his fields nor for biogas production. It is unfit even for making dung cakes. The milk produced in such sheds will also be of poor quality. Since such a milk will have a poor keeping quality, the farmer is obliged to consume it at home rather than sending it for sale. This is a hindrance to the farmer for producing more milk by keeping more and better yielding cows.

Farmers who already have cattle sheds of these types can improve them so that they give more comfort to the animals and permit clean milk production. Availability of money is of course a constraint here. There are ways by which the housing of animals can be improved without spending a lot of money by using wood, mud and straw which the farmers generally possess at home. If one has to spend money on alteration, before spending money the farmer should always calculate what the expenditure will be an what will be the expected yield. If the farmer feels that the expenditure does not yield enough in returns, it is not worthwhile to incur it as he may become poorer and will feel discouraged.

The farmer should make the changes gradually that too by using cheaper and locally available materials. Once the farmer starts producing more milk and earning more money he will be able to afford to spend money to make modem cattle sheds. Some of the suggested alterations for improving the farmers existing cattle sheds are as follows.

1. Surface of the floor in the shed should be made even by leveling off ridges and filling up of pits. The surface can be paved with bricks or at least made hard by good ramming.
2. The farmers cattle shed may be having a roof made of straw or mud or G.I. sheets or asbestos sheets. Straw and mud roofs may leak during the rainy season and harbour insects and pests. If money permits, it is better to replace such roofs with better material.
3. Farmers traditional cattle shed are made very close to or in between living quarters. and the roof is made too low. This makes the shed dark inside and the air movement is also restricted. So as to make it better, the roof should be raised to proper height, at least 6 feet at eaves of the shed.
4. Overcrowding is another problem in traditional sheds. Farmers should take care to see that there is at least about 40 square feet of area per cow and 45 square feet per buffalo in their cattle sheds. This makes the animals existence much more comfortable making them produce better.

Construction Details of Different Structures

The construction of building start from the foundation and gradually rises from plinth to walls and roof. Then different structures are raised inside the buildings. The materials used for construction of different structures depend on their cost, local availability, type of farm buildings and other reasons. The construction details of different structures are described below.

Construction of floor

The floors should be hard, impervious to water and easy to clean. The floor may be of (i) cement concrete or paved with cement concrete flooring tiles, (ii) brick on edge flooring, (iii) stone slab flooring, (iv) *morum* or *kanker* flooring, etc. The best but costliest floors are of cement concrete which may be made only in milking barns, stores, calf pens, shearing rooms etc where strength, cleanliness and imperviousness of floors are important.

In general, cattle sheds floors may be made of stone slabs or brick on edge linings. *Morum* or *kanker* floors are the cheapest but become messy and require constant maintenance.

Floor should be laid over firm plinth, the height of which depends upon the nature of the ground, the location and importance of the building, rainfall, dampness, drainage and possibility of nuisance from insects, scorpions, snakes, etc. its height is normally 45 cm to 90 cm, but can be higher in damper areas. It may be constructed with bricks or stone masonry.

The floors shall have a gradient of 1 in 40 to 1 in 60 towards the drains so that waste water can run into the drains easily. Special care should be taken to make the surface of floors rough and non-slippery in milking barns, passages leading to milking barns, calf pens, etc. Brick on edge floors are sufficiently rough and may not require any special efforts in this direction. The surface of cement floors can be roughened by imprinting the impression ofa piece of expanded metal or suitable wire mesh on the surface while the concrete is still moist.

Construction of walls

Under loose housing system walls are few and not so extensive. The covered area has walls only at the ends. The open area is enclosed by one and a half metres high halfwalls. Walls are built over foundation of sufficient strength.

The walls may be constructed of stone, brick, mud or bamboo or any other material suitable for the locality and the climate. Stone and brick walls are costly but durable and hygienic. Bamboo and mud walls are economical and useful but are temporary and difficult to maintain hygiene. Concrete walls of 10-12 cm thick and reinforced with iron bars along their length and height are the strongest and best but are very expensive. Walls supporting the roof and wall portions with which farm animals come in direct contact must be robust and materials such as bricks, stone or cement concrete may be best for the lower parts of the walls.

Thickness should not exceed 35 cm for ordinary walls, the side walls being fully thick and the lengthwise ones slightly less thick. Partition walls and wall lining the open area should be 22.5cm (two brick) thick. Height of walls should be 2 to 2.5 metres for houses with slopping roofs. Walls and partition can also be made of galvanized corrugated iron or asbestos sheets by fixing to posts 2.5- 3 metres apart. Usually the inner face of the wall is plastered and the outer one painted.

Roof

The roof should be light, strong, durable, weather proof, attractive, a bad conductor of heat and free from tendency to condense moisture inside. There are two types of roofs: slopping and flat The flat roofs are preferred in low rainfall areas, while the slopping roofs are desirable in medium to heavy rainfall areas.

There are several materials available for roofing farm building which includes tiles, slates, G.I. sheets, corrugated asbestos and aluminium sheets, wood, thatch, bamboo etc. G.I. sheets, aluminium and asbestos sheets have many advantages, the chief being fire resistance, casy fixing, durable, hygienic and cheap. These being light material do not require heavy roof supporting structures while tiles and slabes are cumbersone to fix and require heavy supporting structure. Wood make the most comfortable roof, being a good insulator, but is liable to fire risk and is quite costly.

Thatch and bamboo materials are most readily available and are cheap. They are good insulator and can be put over rough and cheap trusses. However, they are unhygienic especially during monsoon and harbour insects, flies, cobwebs, rats, vermins, etc and are highly prone to fire hazards. Though initial investment on thatch roof is less, they are costly in the long run due to the high cost of maintenance and frequent replacement. The slope of roof is expressed as its pitch-angle of the slope with the horizontal. The pitch should be 35 0 for thatch roof, 25 to 300 for a tiled roof and 12 to 180 for a sheet roof. The slope is generally kept steeper in heavy rainfall areas. The pitch should not exceed 450 at any rate.

Mangers

Mangers are very important structures in loose houses as the whole feeding plan may fail because of mangers of improper size. Mangers can be made either in the covered area or in the open lot and should be of continuous type. They can be made of stone slab, wooden planks, bricks lined with cement mortar or of cement concrete. The flooring materials of the manger may be the same as for floor of shed but the surface shall be rounded off and finished smooth. The top of the manger walls shall be arched rather than leaving flat. Mild iron plates may be fixed over the curb (inner wall) of the manger or recessed into curb wall in case the curb is made of reinforced cement concrete. A long pipe of suitable diameter should be fitted all along the length of the manger just above and parallel to the inner wall of the manger. This pipe can be fixed in position by means of a series of angle iron brackets jetting out from the roofs supporting posts. The pipe prevents the animals from getting into the manger or jumping out over the manger and is very necessary in sheds for young animals.

Water troughs

Water troughs may be constructed with reinforced cement concrete, brick with cement mortar, stone slabs with cement jointing or plain thick G.I. sheets. The water tanks may be round (generally in larger pens) or rectangular and shall be located at convenient places in sheds. Its dimensions may be the same as that for mangers for different livestock as per specifications. A two-metre wide paved platform shall be provided adjacent to the water troughs to withstand heavy treading of animals and permit easy cleaning and washing. In smaller pens a tank can be made by raising an impervious partition towards one end of the continous manger.

Hydrants

Each water tank should have one hydrant (a water faucet) the outflow from which may be controlled by a float valve. The hydrant can also be left without a tap and water trough may get filled as the animals empty them. Hydrants with spouts to attach a hose should also be fixed in each shed at convenient places. The pens can be washed by flushing water through the hose.

Drains

Simple open type drains are most suited for loose houses. The drains should be located at about the junction of the covered and open areas. The drains are constructed out of the same materials as for the floor. The width of the drains may vary between 30 -40 cm and a slope of 1 to 40 is provided to the drains. Shallow V -shaped drains with a depth of 6 -8 cm are preferable over drains with cut rides (rectangular or triangular cross section). The drains in a row of pens should be made continous by routing it through holes in intervening partition walls. Outside the buildings, the drain may be led to a receptacle pit if it is desired to store this water. Other wise, the drains can directly be led into the fields where fodders are grown.

Gates and passages

There are two types of gates in a farm one leading into individual pens and the other on farm fences and roads. The latter should be wide enough to allow the passage of trucks, tractors and carts. Similarly, there are passages leading to individual sheds or buildings and main farm roads.

In general, the requirement of service passage are similar, regardless of the kinds of the animals. Specification of gates and passages are given in the table - 3. Gates shall be made of iron or strong wood.

Table 3: Recommended minimum widths for service passages and gates

Sl. No	Kind of passage or gate	Use	Minimum width (m)	Type pf passage or gate
1	Feed passage along the manger	For feed cart	1.5	Metalled
2	Passage leading to individual shed	Movement of animals	1.2	Gravel
3	Farm roads	Truck, tractors	2.5	Metalled
4	Shed gates	To individual pens	1.2	Hinged single
5	Farm gates	Drive through	2.5	Hinged double

Foot bath

A foot bath is a tank measuring 6 x 3 m size at the bottom, 12 x 4 m on top and 0.3 m deep. It is always constructed near the main gate. This tank is filled with a germicidal solution and protects the animals from contagious diseases. Animals and carts entering and leaving the farm have to wade though this solution and in the process the animals feets or the wheels of the vehicles gets disinfected. Thus, no disease producing germs will be brought into the farm through incoming vehicles and animals.

Farm fences

A farm fence marks the boundary, protects the property from intruders, keeps the animals inside, and sometimes serves as a shelter. The fence must be light, strong and durable.

Wooden fences are good for enclosing open lots to animal sheds. The fence can be of post and rail type with four or five 10 x 7 .5 cm cross rails nailed to fence posts, 22cm apart or a pallisade type. Second hand iron rails, angle irons or tubing can also be used for fences. When iron is used, the fence posts should be 2-3 metres apart and fixed in a firm bed of concrete.

Wire fences, however, are more practical and longer lasting. Plain woven wire, wire mesh or barbed wire should be used for fencing. It should be 1.5 m high with 4 -5 stands 0.4 -0.3 metre apart. The line posts of fence should be 3 to 4 metres apart depending upon the type of soil and material of posts.

CLEANING OF ANIMAL SHEDS

The easy and quick method of cleaning animal house is with liberal use of tap water, proper lifting and disposes all of dung and used straw bedding, providing drainage, to the animal house for complete removal of liquid waste and urine. The daily removal of feed and fodder left over in the manger, reduces the fly nuisance. Periodical cleaning of water through eliminates the growth of algae, bacterial and viral contamination and thus keeps the animal healthy.

Sanitation in dairy farm

Sanitation is necessary in the dairy farm houses for eliminations of all micro organisms that are capable of causing disease in the animals. The presence of organisms in the animal shed contaminates the milk produced thus reducing its self life, milk produced in an unclean environment is likely to transmit diseases which affect human health: Dry floorings keeps the houses dry and protects from foot injury. Similarly the presence of flies and other insects in the dairy farm area are not only , disturbs the animals but also spreads deadly diseases to the animals egg. Babesiosis, Theileriosis.

Sanitizers

Sunlight is the most potent and powerful sanitizer which destroy most of the disease producing organism. Disinfection of animal sheds means making these free from disease producing bacteria and is mainly-carried out by sprinkling chemical agents such as bleaching powder, Iodine and lodophor, sodium carbonate, Washing soda, Slaked Lime (Calcium hydroxide), Quick Lime (Calcium oxide) and phenol.

Bleaching powder: This is also called calcium hypo chloride. It contains upto 39 % available chlorine which has high disinfecting activity.

Iodine and lodophor: This is commercially available as lodophores and contains between 1 and 2 % available Iodine which is an effective germicide.

Sodium carbonate: A hot 4 % solution of washing soda is a powerful disinfectant against many viruses and certain bacteria.

Slaked lime and quick lime: White washing with these agents makes the walls of the sheds and the water troughs free from bacteria.

Phenol: Phenol or carbolic acid is very disinfectants which destroy bacteria as well as fungus.

Insecticide: Insecticides are the substances or preparations used for killing insects. In dairy farms, ticks usually hide in cracks and crevices of the walls and mangers. Smaller quantities of insecticide solutions are required for spraying. Liquid insecticides can be applied with a powerful sprayer, hand sprayer, a sponge or brush; commonly used insecticides are BHC, DDT, Gramaxane wettable powders, malathion, Sevin 50 % emulsifying concentration solutions. These are highly poisonous and need to be handled carefully and should not come in contact with food material, drinking, water, milk etc.

Precautions while using disinfection In Insecticide:

- Remove dung and used bedding completely.
- Avoid spilling of dung and used bedding while carrying it out.
- Avoid the use of dirty water in cleaning the sheds.
- Never put the fresh fodder over: the previous day's left over fodder in the manger.
- Prevent algae to grow in the water troughs
- Use proper concentration of disinfectant / insecticide solutions to avoid any toxic effects poisoning.
- Avoid of the mat the milking time as milk absorbs these quickly.

Procedure

- Remove the dung from the floor and urine channel with the help of a shovel and basket (iron) and transfer it to the wheel - barrow.
- Remove the used bedding and leftovers from the mangers in a similar way.
- Empty the water trough and scrape its sides and bottom with the help of a floor brush.
- Wash the water trough with clean water and white wash it with the help of lime mixture once a week.

- Scrape the floor with a brush and broom and wash with water.
- Clean and disinfect the splashes of dung on the side walls, railing and stanchions.
- Remove the cobwebs periodically with the help of a wall brush.
- Sprinkle one of the available disinfecting agents in the following concentration. Bleaching powder should have more than 30% available chlorine. Phenol 1-2% solution. Washing Soda (4% solution).
- Allow adequate sunlight to enter in to the shed.
- Spray insecticides at regular intervals especially during the rainy season (Fly season).
- Whitewash the walls periodically by mixing insecticides init to eliminate ticks and mites living in cracks and crevices.

Other Provisions

The animal sheds should have proper facilities for milking barns, calf pens, calving pens and arrangement for store rooms etc. In each shed, there should be arrangement for feeding manger, drinking area and loafing area. The shed may be cemented or brick paved, but in any case it should be easy to clean. The floor should be rough, so that animals will not slip. The drains in the shed should be shallow and preferably covered with removable tiles. The drain should have a gradient of 1" for every 10" length. The roof may be of corrugated cement sheet, asbestos or brick and rafters. Cement concrete roofing is too expensive. Inside the open unpaved area it is always desirable to plant some good shady trees for excellent protection against direct cold winds in winter and to keep cool in summer.

CONCLUSION

The purpose of cattle housing is to moderate the range of micro-environrnent to which the animals are exposed and optimize their production by protecting them from extremes of climate. The system of housing also determines to a large extent the system of feeding and management of the stock. Therefore, there is a need of providing proper housing to dairy cattle. The climate varies greatly in different regions of the country. For example, the Indo-Gangetic region is largely arid, semi-arid or dry sub-humid and subject to extreme variations in meteorological conditions controlled by seasons. In the Himalayan region, it is temperate, humid to highly humid in the coastal and North-east regions and dry to sub-humid in the Deccan Plateau region. While in northern India, there is distinct winter, in the southern regions there is no winter. Summers are hot every where.

Under such conditions, farmers are adopting diverse housing system in different seasons, such as tethering under trees, tethering close to and under the shade of their dwelling quarters, housing in thatched sheds, enclosing in wood, fenced or mud-walled paddocks, etc. Some better systems of housing such as loose barn housing, stanchion barn housing, etc are also adopted but only by progressive and rich farmers, government dairy farms or commercial dairies. Under these system, the animal gets perfect protection from vagaries of weather from outside intrusion and their best productive capacity is achieved. Though the establishing cost of these housing structure is high, in the long run they prove more economical and fruitful. Therefore, there is urgent need to popularize these housing systems with desirable modifications among the farmers.

Loose Housing Construction

Loose housing maybe defined as a system where animals are kept loose except milking and at the time of treatment. The system is most economical. Some features of loose housing system are as follows.

1. Cost of construction is significantly lower than conventional type.
2. It is possible to make further expansion without much change.
3. Facilitate easy detection of animals in heat
4. Animals feel free and therefore, prove more profitable with even minimum grazing.
5. Animals get optimum exercise which is extremely important for better health and production
6. Overall better management can be rendered.

The floor and manger space requirement of dairy cows are give in Table below

Other Provisions

The animal sheds should have proper facilities for milking barns, calf pens/ calving pens and arrangement for store rooms etc. In each shed, there should be arrangement for feeding, manger, drinking area and loafing area.

The shed may be cemented or brick paved, but in any case it should be easy to clean. The floor should be rough, so that animals will not slip. "The drains in the shed should be shallow and preferably covered with removable tiles. The drain should have a gradient of 1" for every 10' length. The roof may be of corrugated cement sheet, asbestos or brick and rafters. Cement concrete roofing are too expensive.

Inside the open unpaved area it is always desirable to plant some good shady trees for excellent protection against direct cold winds in winter and to keep cool in summer.

Type of Animal	Floor space per animal (Sq. feet)		Manger length per animal (inches)
	Covered Area	Open Area	
Cows	20 – 30	80 – 100	20-24
Buffaloes	25-35	80-100	24 -30
Young stock	15-20	50-60	15 -20
Pregnant Cows	100-120	180 – 200	24 -30
Bulls Pen	120-140	200-250	24 -30

The Floor Covering

The selection of the floor covering of the housing area proper and the surrounding space (feeding passage, holding corridor) is very important. On the one hand, the floor covering is an important item in the cost price of construction. On the other, proper utilization of livestock housing depends somewhat on what kind of flooring is selected.

In terms of cost, the range of difference between a floor which is merely tamped down and one where concrete is poured can vary from 1 to 20.

In terms of utilization, improper flooring can impede the movements of both man and beast, aggravate the hardship of the actual labour involved, and jeopardize the animals' cleanliness, comfort, and even health (lameness, various kinds of accidents…)

The major possible alternatives are:

- Ground left as is, perhaps merely well-tamped
- Ground partly or wholly cobbled and flattened with some mechanical device;
- Ground partly or wholly floored with concrete;
- The use of slatted flooring is not envisioned, we remind the reader, because of its excessively high cost and the complexity of handling and storing manure with this system.

Floor covering, therefore, must be selected on the basis of the following constraints:

Climate — particularly rainfall intensity and frequency;

Soil type over which housing is built: texture, slope and corresponding drainage capacity;

The way the housing is used - one can be much less demanding with regard to floor coverings for housing utilized only during the dry season than with year-round housing;

Animal density: as we have seen, density can vary, with the full open yard, for instance, in proportions of 1 to 3 according to the drainage capacity of the soil.

The housing system: paradoxically, it is the fully roofed or combined systems where flooring is less of a problem, because here there is protection from the rain.

The individual sections of the housing: floor covering is a different problem according to which of the three already discussed areas (lying, feeding, exercise), or surrounding areas, it applies to.

Fully open yard

It has been repeatedly emphasized that fully open yard housing should only be built on proper, meaning properly-drained, soils.

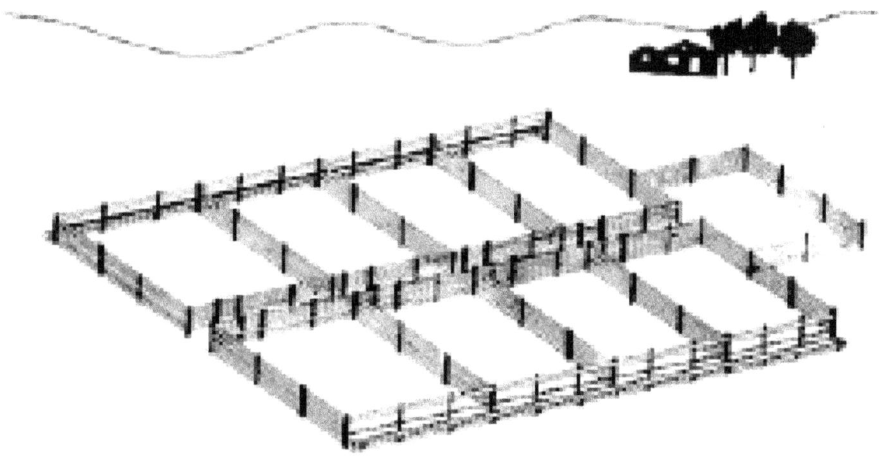

Fig : Fully open yard

The economic interest of this system lies precisely in the fact that it needs neither roofing nor flooring. The ground may, however, sometimes be treated differently in the three sections of the housing:

Rest and exercise areas: The ground will be left as is or carefully drained so as to increase its resistance to trampling by the animals' hooves. Indeed, in view of the area allotted to each animal, better flooring such as blocking or paving would drive costs up to the point where the system would no longer be competitive with roofed systems.

Feeding area: this area of the housing, a strip 2–3 m wide along the manger, is usually treated the same way as the rest and exercise areas.

Since the animals drop most of their dung in the feeding area, it might be a good idea to treat this area somewhat differently, such as blocking and then flattening it with some mechanical device, for example, or even pouring concrete.

In the latter case, the sidewalk system is used: a horizontal platform 2–2.5 m wide is built along the manger. From time to time, the dung and refuse which have accumulated on the sidewalk are cleaned with a tractor rear- or front-mounted dung scraper, taken out of the yard and dropped into a slatted trench where the liquid manure can filter off, resulting in a dry manure which is much easier to handle and transport.

THE COMBINED SYSTEM

The amount of ground area is much smaller in the combined than in the preceding system (1.5–2 m2 roofed and 2 m2 open). Here the lying area is under the roof and the exercise and feeding areas comprise a single section in the unroofed portion.

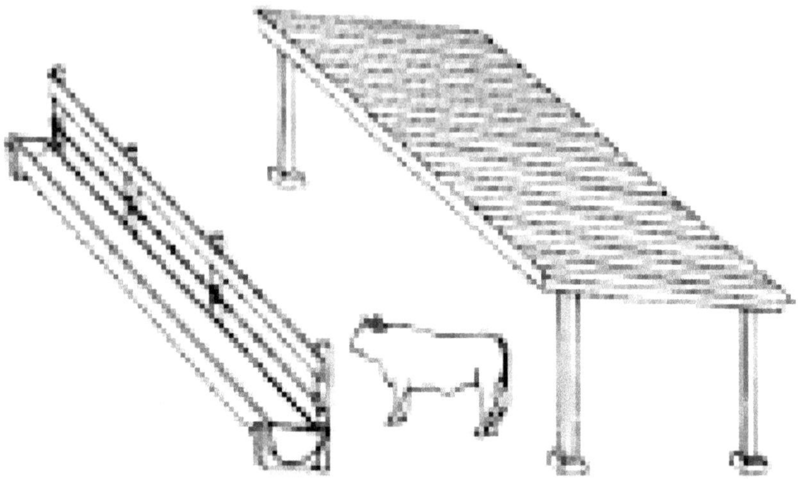

Fig : Combined System

Covered area: As it is not subjected to run-off, there is no point in giving the ground in the lying area any special treatment. The original levelling and tamping of the soil is sufficient, with the proper litter, to ensure the animals' comfort.

Open-air section: This is the feeding and exercise area. This is also the section of the housing where the tractor and trailer cross during manure removal. And lastly, this is the area *where most of the dung collects*.

If rainwater does not run off this section of the yard, it can become a real cesspool where the animals will have a hard time moving about, especially during the rainy season.

So the ground in the open part of the combined system should not be left as is, or simply tamped as in the preceding system: there will have to be some sort of treatment, which might include:

- Blocking followed by rolling, where drainage is satisfactory;
- Concrete flooring in this section where drainage is poorer; this section can be cleaned with a tractor dung scraper.

Unfortunately, we often see stockmen doing just the opposite; proceeding to pave the sheltered area of combined housing whilst leaving the ground in the open section as is. For the same expenditure, the housing loses much of its effectiveness due to the fact that the concrete is not serving any purpose and the feeding area becomes impassable during the rainy season.

Full Roofing

The housing is entirely covered by a roof, so it is protected from rain and runoff. However, given the high density of animals/ unit of area, different solutions can be envisioned:

Floor covering: depending on the availability of litter (straw, wood shavings, or other absorbant, bio-degradable and inexpensive materials).

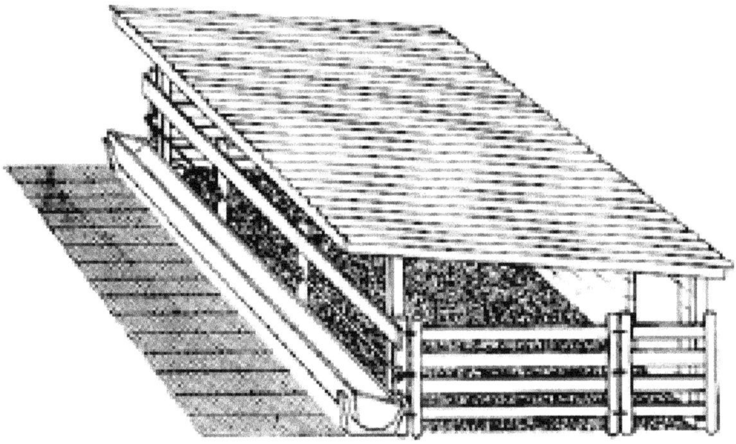

Fig : Full Roofing

When litter is available and/or manure has some value, the housing floor can be treated in the same way as the covered part of the combined system. In other words, simply levelled and then covered with deep litter. Here the need for proper amounts of straw (2 kg/head/day) must be re-emphasized to ensure absorbtion of the moisture from urine and dung and hence the animals' cleanliness and comfort.

Moreover, where the animals are watered with an automatic drinking-bowl, it is a good idea to pave a one-meter wide area all around the bowl.

On the other hand, where litter is scarce (straw used as feed or too expensive), *a sloping concrete platform will be built*, (with a gradient of 8o in the direction of the manger) and there will be a semi-liquid manure drainage channel running along the feeding passage. With this system, the housing is automatically cleaned by itself as the animals trample the manure into the channel.

Flooring of annexe areas: (observations valid for all three housing systems).

The feeding passage has to be used several times a day by workers using a trailer (preferably animal-drawn). So this passage must be passable at all times. The ground will have to be blocked and rolled, cobbled, or, the more economical solution, asphalted.

When the housing is set up on a concrete platform, a rather wide cement channel (at least 30 cm wide and deep) should be installed immediately in front of the manger so as to allow a worker to clean out the dung trampled into the channel by the cattle. This cleaning is done from time to time.

The holding and circulation passage: is only used by the cattle on rare occasions — at the end of the breeding/fattening cycle and perhaps during periodic weighings. So nothing has to be spent on treating the ground in this passage; it can be left as is.

The Manger

The manger fulfills a dual function in the housing systems described in this report: as a feeding area and as a partition separating the cattle from the outside. Special care must therefore be taken with its construction.

Its dual function means that the manger is one component of the housing which receives the most abuse from the animals as they rush onto it when the concentrates are distributed. It is therefore a good idea to overdimension the materials used so that the manger will not be quickly wrecked or damaged.

It should also be noted that the design and size of the manger influence the ease with which the animals absorb the fodder they receive. Independently of waste (fodder which has been refused or trampled upon after falling from the trough), the way the manger is built has a direct influence on the animals' daily consumption and hence on the technical and economic consequences of the operation.

Because of this, no matter what materials are used it is imperative to respect those standard dimensions which have been worked out after lengthy trials. These refer especially to the diameter of the trough, its height above ground and the space between it and the withers bars.

By the term "manger" we include the entire feeding-trough apparatus together with its *protection devices* such as the withers and protection bars which keep the animals out of the trough.

Various materials can be used in the construction of the trough:

Wood and iron, with which movable troughs can be built;

Reinforced concrete and parpen;

Semi-circular *irrigation channel pipes* where available.

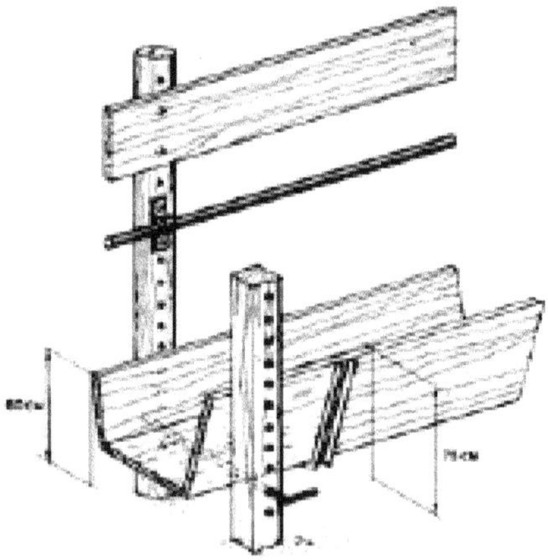

1. The different materials of the trough

a) The wooden or sheet iron trough

Economic reasons may determine *the choice of one of these two materials*, though they can hardly ever compete with the other solutions suggested, in order to increase the *flexibility of utilization of the building*, which may have been remodelled in such a way that the easiest access is by movable troughs, but also for *specific technical reasons*, since these materials are light enough to allow the installation of a rack-bar to adjust the vertical height of the manger.

Wood is the easiest material to work with, but it is not available everywhere at acceptable prices. Another factor is that wood mangers usually have a shorter useful life than do mangers built of other materials.

Iron is also a rather easy material to work with, but it needs more equipment to work with (welding equipment) than does wood. And its cost is often excessively high in the developing countries. And iron mangers are often attacked by certain rather acid feedstuffs, such as silage, and by moisture (unless the iron is galvanized, which ups the cost further). Nor is their useful life necessarily longer than that of wooden mangers.

The construction of wooden or sheet-iron mangers is subject to the afore-mentioned building standards. One necessary component is a *standard protection device*, no matter which material is used.

The basic advantage of sheet-iron or wooden mangers is that they are lightweight — elements up to 6–8 meters can be handled easily. This means *a rack-bar system can easily be installed* so that height can be regulated in terms of the ground level of the housing. By the time the breeding/feeding cycle is over in a fully-roofed, deep litter, small space housing arrangement, the ground level may have risen over 50 cm. This rack-bar can be made from two metal bars which are slid under the trough into slots or holes in the supporting posts and the ends of the trough.

The use of this trough is particularly recommended for fully roofed loose housing, where the trough is protected from inclement weather and can be raised in height as the litter accumulates. Its use may also sometimes be recommended in combined housing using remodelled buildings where lateral access is impossible. Should this be the case, the trough can simply be taken down to allow dung removal or entrance or exit of the animals.

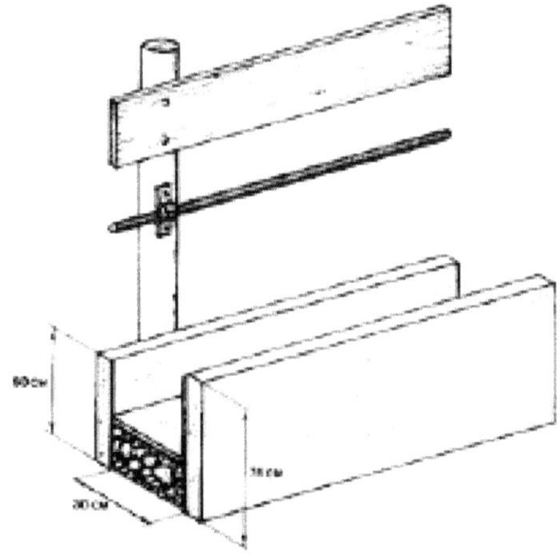

(b) The parpen (breeze-block) or reinforced concrete manger

These materials do not always mean either the most *economical* or the most *flexible* solution. In addition to the fact that concrete and breeze-block managers can neither be moved or adjusted in height, they also present the *disadvantage* of preventing the installation of certain manure removal systems (such as the sloping concrete platform). Still, if breeze-block and reinforced concrete often meet with the approval of livestock raisers, it is because they can usually be used by *fairly unskilled labour*, and especially because the installation is a very *sturdy* one which can be amortized over a ten to twenty year period.

Parpen (or conglomerate) is easily manufactured on the farm with a minimum of equipment and at relatively low cost. However, it is not always resistant enough (this depends on the composition of the mortar used in making them).

Reinforced concrete, a relatively easy material to work, has almost unlimited resistance. But its cost price is often considered excessively high.

The parpen or reinforced concrete manger is built by making two parallel partitions 60 cm from one another. Parpen partitions can be 15 cm thick and reinforced concrete 8–10 cm.

The partition on the side of the housing is 60 cm high. The partition on the side of the feeding passage is 75 cm high. In this manner the cattle are unable to push the fodder outside the trough (and outside the yard) with their muzzles. The area between the two partitions is filled with pebbles up to a height of 30 cm, leaving a trough depth of 30 cm on the side where the cattle feed.

After plastering the whole arrangement with a mortar rich in cement, the *standard protection device* will be installed as per the other manger models.

Wherever semi-circular irrigation pipes are unobtainable, the use of this kind of trough is recommended for fully open yard and combined housing systems. As a matter of fact, since this kind of trough is attached to the ground by foundations, the masonry trough can obstruct the passage of dung towards the feeding passage. This would make it difficult to set up a sloping concrete platform, which is of interest in areas where straw is too scarce to enable the deep litter system to be used.

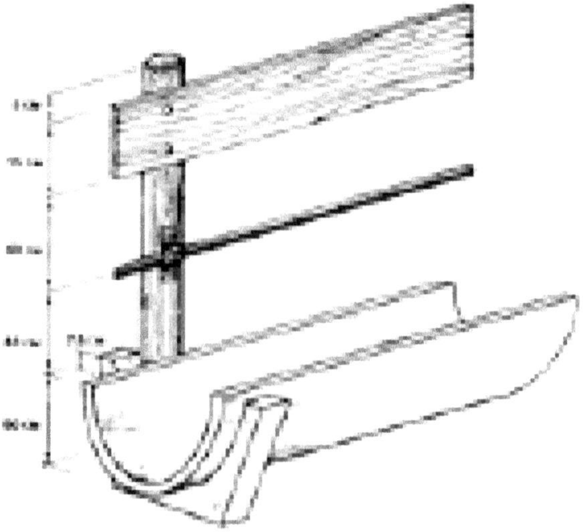

(c) The irrigation pipe trough

This is the recommended material for use when the farm is at reasonable distances from where the pipe is manufactured.

These semi-circular pipes of vibrated concrete have an interior diameter of 60–80 cm. Trapezoidal pipes, with an interior diameter of less than 60 cm have proven less resistant and are to be avoided, as they do not offer sufficient capacity for the distribution of voluminous feeds such as fooder (silage, hay, green fodder).

This is *the simplest* and quickest kind of manger to build. All the pipes are laid onto their prepoured supports at the same time, thus cutting down on labour supervision.

It is also a *very sturdy material*, resistant to the shocks of bad weather as well as the weight of the animals. Its useful life is virtually unlimited. In the final analysis, this is the *most economical solution*, especially when the pipes can be bought second hand or at bargain prices (should they not be absolutely waterproof, and thus unusable for transporting water, this defect has no effect upon their use for feeding-troughs).

A detailed description of *how to build feeding-troughs* from irrigation pipes is given in the following pages (pps. 74 to 83). Let us just mention at the outset that use of this material is limited only by transport and handling problems (where the pipes are available).

The weight of each component, measuring 6–8 m in length, is such that they have to be moved to the building-site by large trailers, and these cannot always get through the trails leading to the building-site.

Unloading should preferably be mechanical, so a crane ought to already be available on the site.

In light of the costs involved in moving the pipe, its use would not seem to be recommendable unless the building-site is less than 100–200 kms from where these pipes are available.

The semi-circular irrigation pipe trough can be recommended for use in all types of housing. In any case, to the extent this trough is too heavy to be able to be adjustable in height by means of a rack-bar, its use will preferably be reserved for fully open and combined housing systems.

The existence of a space of about 30 cm beneath the trough and the ground means that irrigation pipes can be used with concrete platforms which are so useful in areas where straw is scarce.

2. The various components of the manger

Though the manager may well be built with various materials, the authors have decided to provide a detailed description of a manager, its protective devices, and assembly, based on the use of irrigation pipes. This does seem to be the material with the best cost/strength ratio in most cases. And it is also a rather original technique, which deserves an explanation.

Concerning the protection system (pps. 78 to 83), this is identical, whatever material is used in building the manger.

(a) The supports for the irrigation pipes

Various materials can be used to support the irrigation pipes. The manufacturers of these pipes often deliver them with vibrated concrete supports designed to fit the outer sides of the pipes. But when such supports are not available, they must be built from *bricks or breeze-blocks*. The use of such lightweight materials as wood and metal is not recommended as they are too weak to support the weight of the pipes.

Where the supports are delivered with the pipes, *they must be bedded* in accordance with the following standards:

Distance between the axises of the supports must be exactly equal to the length of the pipes used so that where battery housing is used, the end of each pipe can be laid on one-half of the support (the other half is for the next pipe). This material is so strong that no intermediate support is needed.

Support thickness varies with the material used: about 15 cm is enough for reinforced concrete. A row of bricks where brick is used.

Support height is basic as the height of the trough is determined by it. To respect the height of the trough, the lowest part in contact with the pipe should be about 20–30 cm off the ground, according to whether the pipe measures 60 or 80 cm, respectively, in diameter, and about 7 cm thick.

Horizontal supports on sloping ground would not be a desirable element; the essential priority is the height of the edge of the manger with respect to the ground. In the no-roof, fully open yard system, the manager will be set up on the highest part of the site so as to avoid the accumulation of water in the feeding area, which is where most of the excreta accumulates.

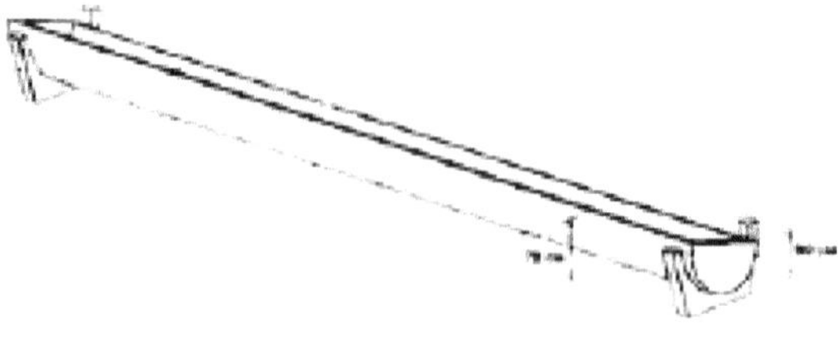

(b) The manger proper

As has been emphasized, there are many advantages to the use of irrigation pipes for feeding-troughs. But these pipes must be used in accordance with a number of highly specific recommendations.

The pipe is laid asymmetrically on its supports. The two edges of the pipe are not level with respect to the ground. The side of the pipe facing the housing has to be about 15 cm lower than the side facing the feeding passage.

It is this asymmetry which prevents the animals from knocking their fodder over the edge of the trough with their muzzles.

Where very voluminous feed is distributed (hay, straw), the asymmetry of the trough can be added to by fastening a plank about 20 cm wide in a vertical position above the outer edge of the trough.

The pipe has to be laid at the proper height with respect to the ground: the right height is 60 cm aboveground for the side the animals are on and 75 cm for the side facing the feeding passage.

How accessible the feed is determined by the height of the manger on the animals' side of the trough. Obviously, what is needed is an *average height* applicable to both 100 kg cattle housed immediately after weaning and 450 kg young cattle ready to leave the housing — slaughter for the males and breeding for the females.

It is imperative to respect this standard. Anything higher would block the animals (especially the younger ones) at dewlap height, and would keep them from getting to the bottom of the trough — to the concentrates, for instance.

Anything lower would be wrong because the animals grow in size and the ground level rises where deep litter is used. The result would be that the manger would be filthied, the animals would have to kneel to feed, and there might even be accidents as the animals, with their feet in the trough, were jostled by the others.

Stockmen often build their mangers level without taking into account the slope of the site. The result is that the edge of the manger is the wrong height along much of the housing. If a livestock raiser is bothered by having a sloping manger, then he will have to agree to the outlay for terracing prior to construction of the manger.

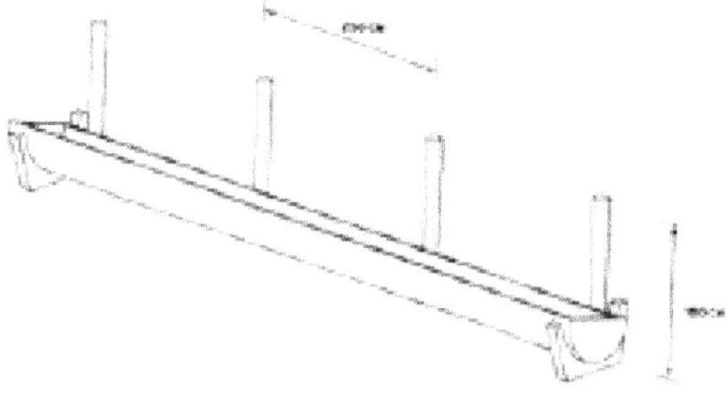

(c) Support posts for the withers bar

Various materials can be used for making the posts to support the manger protection system (withers and protection bars):

The use of *wood* makes it considerably easy to fasten on the elements the post is intended to support. And it is usually cheaper. Hardwood sapling poles 15 cm in diameter are preferable to sawed-off pine wood which has to be 20 cm through. In any case, the posts will presumably be treated to prevent rot.

The more aesthetic *metal tube*, with an external diameter of 80 mm, can also be used. But it is more expensive and is more difficult to attach the withers and protection bars to.

The *reinforced concrete post* may sometimes be the most economical, but, here again, there will be problems with attaching the protection bars to it.

The *dimensions of these posts* vary according to the bedding system selected, no matter what material is used.

The above-ground measurement of the post is that of the fences: 1.60 m. If the posts are bedded with road metal and coarse concrete, they can be sunk to a depth of just 50 cm, making a total length of 2.10 m.

However, if, for economic reasons, one should decide to not use cement and use only tamped earth to hold the posts, they should be sunk to a depth of 80–100 cm, or a total length of 2.40–2.60 m.

Bedding the posts:Spacing between two successive posts must not be greater than 2.50 m. The substantial pressure put on the withers bar by the cattle when the feed is distributed would mean that any greater spacing than 2.50 m would make the protection device too fragile.

The posts are so arranged as to touch the top side of the trough on the side of the housing: if the posts were placed away from the trough, the withers bar would keep the cattle too far away from the trough.

And lastly, for prolonged use of the installation, it would be desirable to *protect the posts* from attack by urine and litter, especially where the deep litter system is used. This is particularly important when metal posts are used. Set into a block of concrete about 30 cm above ground level, they will be sufficiently protected.

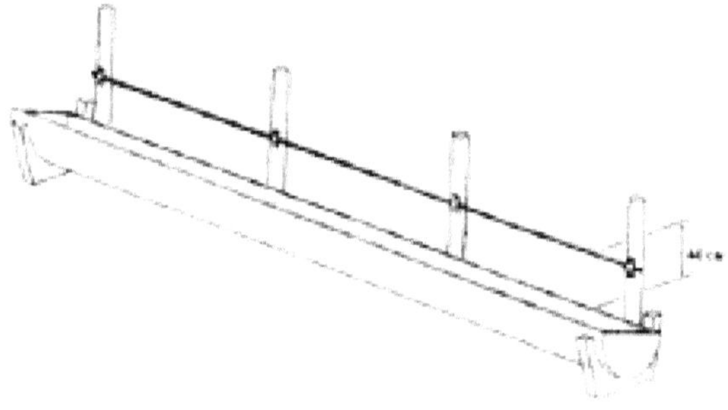

(d) The withers bar

The withers bar has an essential function. This is a horizontal bar which, as its name indicates, is located at the animals' shoulder height so as to keep them from getting too far into the manager whilst at the same time allowing easy access. With the withers bar, it is at the same time possible to:

- Prevent the cattle from *getting their feet into the trough;*
- Force them to *feed perpendicular to the trough* and therefore reduce to the minimum the trough width occupied by the individual animal;

- Keep the animals from *getting out of the housing*. We might say the manger fulfills the dual function of providing food and fencing the cattle.

Materials suitable four use for withers bars have to be sturdy enough to resist the shoving action of all the animals rushing at once to get to the trough to eat their favourite foods, such as concentrates.

The ideal withers bar is the *metal pipe* with an internal diameter of 50 mm and an external diameter of 60 mm. However, if this is not available, a smooth, cylindrical *wooden bar*, with a diameter of at least 120 mm, can be used.

It is not advisable to use a *board*. It would be too fragile, considering the distance between the posts.

The bar is fastened on the side away from the housing. In other words, over the manger. Though it would seem that to fasten the bar on the animals' side would make it more solid, this is to be avoided as it would push the animals back from the manger at a distance equalling the sum of the diameters of the post and of the bar.

The bar is fastened in the same way for a movable or stationary manger. A metal bar will be fastened by two forged hasps bolted onto the post. A wooden bar will be fastened by two bolts going through both the post and the bar. For movable mangers, regularly spaced holes will be drilled into the posts (e.g., every 10 cm), so that the withers bar can be raised at the same time as the manger is moved.

The distance between the upper edge of the trough on the side of the animals and the withers bar must be 40 cm. For animals weighing under 200 kg., it can be cut to 35 cm and for animals weighing over 500 kg it should be increased to 45 cm.

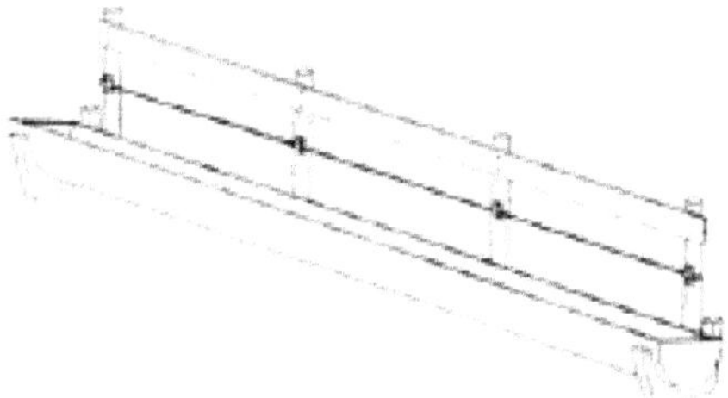

(e) The protection bar

The trough and the withers bar would not be enough to keep the young bulls from climbing over the manger without a protection bar above them. *In the final analysis, it fulfills the function of being the highest part of the permanent fence.*

One can, in fact, given the aforementioned measurements, calculate the height of the withers bar at a bit more than 1 m off the ground, which would not be enough to keep the animals from jumping out of the yard.

It is rare that the protection bar has to bear the weight of the cattle. Its part is rather to dissuade them from trying to jump or climb over the manger. *The materials used to make it*, therefore, are more lightweight. They will be the same as those used in the permanent fencing: 35 mm boards, 100 mm diameter wooden bars or 500 mm diameter metal tubes.

They are *fastened in the same way*; for the pipe, and for the board or wooden bar.

The protection bar is fastened so that the height of its upper edge is the same as the top height of the permanent fencing - 1.60 mm from the ground.

To conclude this chapter on the manger, the following items must be emphasized:

- Each of the components described (manger, posts, withers and protection bars) has a *highly specific and irreplaceable function*;
- *The choice of materials and how they are fastened is rather flexible*. The only decisive criteria, other than the economic factor, is their resistance to prolonged, intensive use. Looking at it from this standpoint, compatible materials (meaning, having the same useful life) have to be used. It would be silly to build a manger which could last for 20 years and a protection system which had to be replaced after one or two.
- *The dimensions* (especially trough and withers bar heights), are absolute imperatives. They have been formulated after many and extensive trials. To modify them in any way would be to jeopardize the best possible return from all inputs.

THE DRINKING-BOWL

Young cattle need a permanent, always accessible source of water. Proper watering is an *a priori* in successful cattle raising.

The poor results which have sometimes come out of conventional stanchion stable establishments which are not equipped for frequent watering of the animals have to do with the fact that insufficient watering brings about diminished fodder consumption and therefore a lower performance platform.

Among those factors influencing the daily intake of water are liveweight, the kind of fodder distributed, dry matter content of the cattle's diet, individual performance, climatic conditions, how often the animals are watered, etc.... It can be estimated that the average daily water intake of young cattle weighing in at 300 kg is about 30 litres, which varies substantially with climate and diet.

A single manger model has been presented for all types of housing and materials used. However, different solutions can be envisioned for the drinking bowl:

When *the farm does not have a supply of running water*, the best plan is to store enough water in the housing area to supply the animals' needs for over 24 hours. The best solution in this instance is the large-capacity reservoir tank. But unfortunately it is only suitable for fully roofed open housing.

However, *when the farm does have running water*, the reservoir can be a rather small one if it refills automatically.

The construction of any type of loose housing which would not have a permanent supply of water accessible at all times to the animals is definitely not recommended. To take the animals out of the enclosure once or twice a day to drink would inevitably reflect unfavourably on their performance.

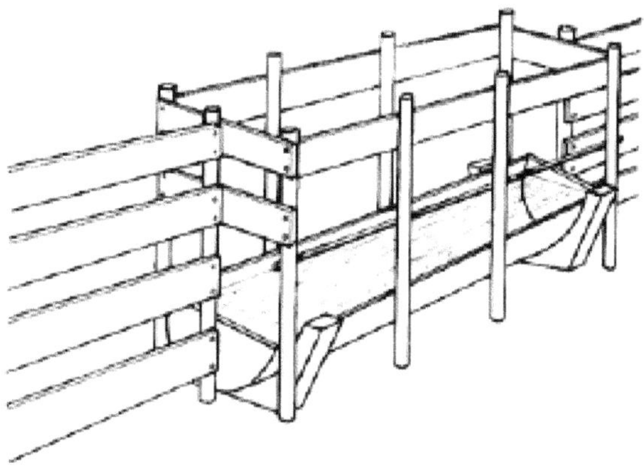

1. Watering-trough reservoir

Supplying water: The reservoir-watering trough will be installed only on farms without running water. Indeed, the much higher installation and operating cost of this system is only justified where water has to be transported by movable tanks from a rather distant source of water.

A unit of more than 20–30 animals presupposes the installation of a reservoir watering-trough which would require the permanent presence on the farm of a tank trailer and tractor. For smaller units, animal traction is recommended.

Description of the watering-trough: A reservoir-type drinking-trough is made up of a tank with a 50-litre capacity per animal and the accessory protection devices. *The simplest solution is a semi-circular pipe* identical to the one used for the manger. It must be closed at both ends by a cement partition and laid horizontally with both edges at the same distance from the ground. A pipe 60 cm in diameter and 8 metres long has a 1 100 litre capacity and can meet the needs of a twenty-head lot. A pipe with an 80 cm diameter (internal diameter in both cases) can be used for larger lots. For the same 8 metre length it can contain about 2 000 litres, which can water a lot of nearly 40 head.

If these pipes are not available, *the tank can be made of breeze-block or formed concrete:* the top of the tank should be 60 cm off the ground, the depth 30–40 cm, and the width 60 cm. These masonry tanks have a capacity of 180–240 litres/linear metre.

The use of metal tanks is only justifiable economically speaking if the tanks were purchased for some other purpose and are simply being adapted for use as watering-troughs.

Whatever material is used in making the tank, it should receive the same protective devices as the trough; i.e., withers and protection bar.

One possible arrangement for the tank/trough is to build it between two lots, whereupon it becomes part of the partition separating two adjacent yards. In this case the protection system has to be set up on both sides of the watering-tank.

The tank/trough is very adaptable to no-roof open yard systems, and, where strictly necessary, to combination-type housing. In the latter case, it has to be installed under the roofed section, which limits the length to 3–4 m (and hence the capacity).

The installation of the tank/trough in fully roofed loose housing is not generally recommended.

Finally, it should be noted that carrying water by tank trailer weighs heavily on the budget for feeders/breeders: it is always desirable. to examine the possibility of piping water and thus making possible the installation of one of the two above described drinking-bowls.

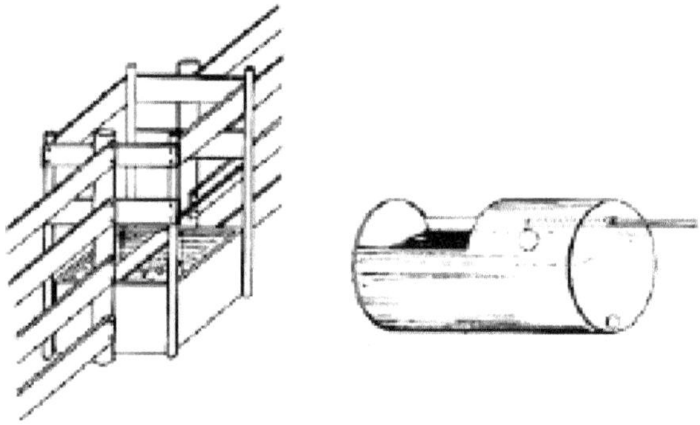

2. Constant-level drinking-bowls

Water supply: Wherever possible, the installation of water under pressure is recommended for young cattle housing

- *Plugged into an existing network of pipes;* of course the choice of the housing site will have to take into account such a network and its location;
- *Constructing a water tower* with mechanical lifting such as an electric, motor, gasoline or wind-powered pump.
- *Setting up a reservoir* at a higher elevation to provide water for the housing even though refurbished by tank truck, when the spring or well is far away.

It does seem excessively costly to set up a supply of water under pressure, but in actual fact it always translates into considerable savings when compared with watering from tank/troughs, because:

- The automatic supply is a labour-saving device;
- Water under pressure removes the risk of a water shortage, which can result in lowered performance;
- With water under pressure, the smaller drinking-bowls, which are better adapted to combination and fully-roofed loose housing, can be utilized.

Description of the watering-tank:

A 100–200 litre tank can be made of concrete and designed to include protection and withers bars so that the animals cannot get into it. The amount of water entering the tank is controlled by the sort of ball found in a flushing toilet and hidden beneath a sturdy metal cover. The level of the water can thus be maintained constant.

A tank of this type can also be devised much more simply and less expensively from a 200-litre metal tank. A portion of the metal is removed so that the animals can drink from the inside of the tank, and the ball arrangement is protected by the covered portion.

The use of the constant-level drinking-bowl is recommended for all types of housing at the rate of a single apparatus per lot in the case of a metal tank and one apparatus shared between two lots in the case of a concrete watering-trough.

Indeed, the constant-level drinking-trough of local manufacture can usually be an advantageous substitution for the automatic drinking-bowl in most of the developing countries.

3. The automatic drinking-bowl

Water supply: The advantages of installing water under pressure in feeder/breeder installations can hardly be overemphasized. The characteristics of the automatic drinking-bowl, which only has a capacity of about 1 to 2 litres, and which refills as the animal drinks, mean that a certain minimum of pressure is needed for correct utilization. So where pressure is low (such as a reservoir of water at not much greater elevation than the housing), this system is to be avoided.

Description of the drinking-bowl: Usually made of cast-iron and delivered ready for immediate use, it has only to be connected up to the running water supply. It has the advantage of taking up very little room. Unfortunately, it cannot usually be manufactured in the developing countries and must be imported at relatively high cost.

The automatic drinking-bowl is of universal application. It fits all types of housing and is especially suitable for the fully roofed loose housing system because it takes up so little room. One drinking-bowl can accommodate 15–20 head (one at a time, of course), and hence lots housed in combination or fully roofed loose housing.

The location of the drinking-bowl, whether automatic or constant-level systems are used, must conform to the following recommendations:

Fully roofed open housing. The drinking-bowl will be installed in the exercise area, along the stationary side fences.

Combination-type housing. It will be installed just immediately before the roofed area.

Completely open yard housing. The drinking-bowl will be installed in one of the rear corners of the yard.

No matter which type drinking-bowl is selected, it must under no circumstances be installed in the centre of the housing (which would hinder mechanized litter removal), by the movable fences (which would prevent access to farm machinery) and near the manger (hindering access to the feeding-trough).

THE ACTUAL HOUSING

Here we wish to discuss all those superstructures which fulfil the functions of protection (roof and protective wall) and partitioning (stationary and movable fences).

In choosing the materials for building these elements, the builder ought to demonstrate as much imaginativeness and flexibility as possible.

Indeed, the list of materials given here is by no means exhaustive. And this is the area in which the boldest and most economical solutions can be sought.

Common sense and cost should be the only guidelines dictating the choice of materials. In any case, it is helpful to keep in mind *the function* of each component and consider only those factors connected with fulfilling this function.

The function of the roof, where there is a roof, is only to protect the lying area from too much sunshine and rain. The kinds of conventional (traditional) roofing already found in most of the developing countries are quite adequate for this purpose.

The function of the walled-off area, where there is one, is only to protect the lying area from the action of the prevailing winds. It is easy to imagine that this being the case, what is needed is not necessarily a wall but a sort of screen, which can be made from the most diverse materials.

The function of the stationary fence, as has already been mentioned, is to prevent the animals from getting out of the housing. In this area, each country has its own customs, and these can eventually be used to great advantage.

The function of the movable fence, is the same as that of the stationary fence. However, it must be able to be opened easily and closed firmly. Experience appears to demonstrate that movable bars are preferable to the conventional stockyard gate.

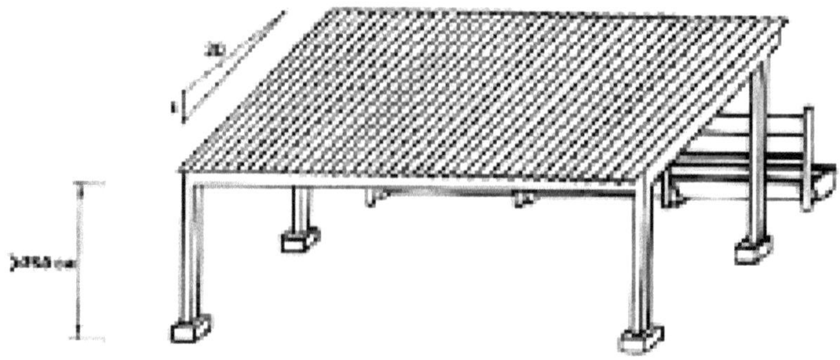

1. Roofing

Roofing construction methods for combined and fully roofed housing are flexible for what concerns the materials to be used. But concerning the actual construction of the roofing, methods must be more strictly adhered to.

The materials which can be used are extremely varied. In roofed housing, this is the component with the highest cost per head. So the solution opted for must offer the best cost/strength ratio - meaning the one involving the lowest possible amortization cost.

The frame can be made out of: *wood* (easy upkeep and relatively resistant), iron (minimal upkeep but expensive), reinforced concrete (very resistant but not reusable should the building be remodelled for some other use). These components can, eventually, be prefabricated.

The roof can be made from *industrial tile* (the disadvantage is that it is heavy and easily broken); from *eternit* (light and easy to assemble but fragile): from *sheet metal* (even more lightweight but rusts easily). Conventional (traditional) materials such as puddled clay and thatch can also be used. Though they do occasionally have to be redone, there is no initial outlay for materials and they are excellent insulators from heat.

The construction of the roof must respect a certain number of essential specifications:

The *roof frame* must be high enough to allow for normal circulation of men and animals, and possibly such farm machinery as the forklift, and it must be high enough after the deep litter has piled up (it can pile up as much as 50 cm). The lowest point of the roof frame must be more than 2.5 metres off the ground. This recommendation is of particular importance in the fully roofed type of open housing, where tractors have to be able to manoeuvre at some point.

Since the depth of the fully roofed housing is 6–8 m and the slope is 20 percent the height of the edge of the shed on the manger side is about 4 metres.

The slope of *the roof* must, in fact, always be oriented towards the rear of the building. There is no problem with the accumulation of water in the circulation and holding passage, since the passage is only used by the cattle once a month.

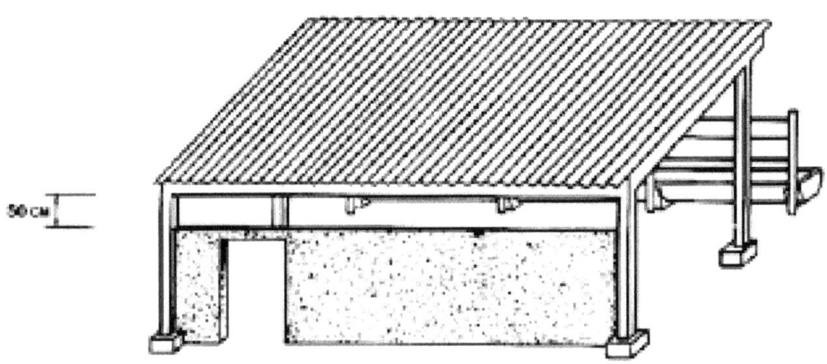

The roof must never be oriented towards the front. Because in the combined system, the rainwater would then accumulate in the exercise and feeding area, and in the fully roofed yard it would accumulate in the feeding passage, making them both impassable.

2. The shelter wall

Like the roof, the shelter wall is only found in the combination and fully roofed yards, which have to be oriented against the direction of the prevailing winds. In order to protect the animals, the end of the covered part is boarded up (or other materials are used) to provide a wind screen.

The materials which can be used for the shelter wall are also very varied.

The most satisfactory solution is to build a brick, breeze-block or stone wall, but it is also the costliest one. It is cheaper to use sheet-metal, boards, and even thick bales of straw protected by fine wire netting on the side of the animals, even if it has to be reinforced by a stationary fence. Nor ought one to eliminate such traditional solutions as the puddled clay (adobe) wall, which involves no other expense than the cost of labour.

Whichever material is selected, the shelter wall must include a ventilation system and an entrance/exit door.

So as to ensure that the housing will be well-aired, it is a good plan to leave a free space about 50 cm high on the top part of the rear wall and all along the building. This opening, which lets the hot air circulate in the summer, has to be able to be closed off in the winter by windows or even by bales of straw or just by fairly strong sheets of plastic (plastic sacks).

We remind the reader that an entrance/exit door allowing access to the housing to both people and cattle has to be built in for each lot. This door is set in the protective wall, and provides direct access to the circulation and holding corridor located behind the building. The door is set into a corner to facilitate the handling of the animals and should measure 1 m wide by 2 m high.

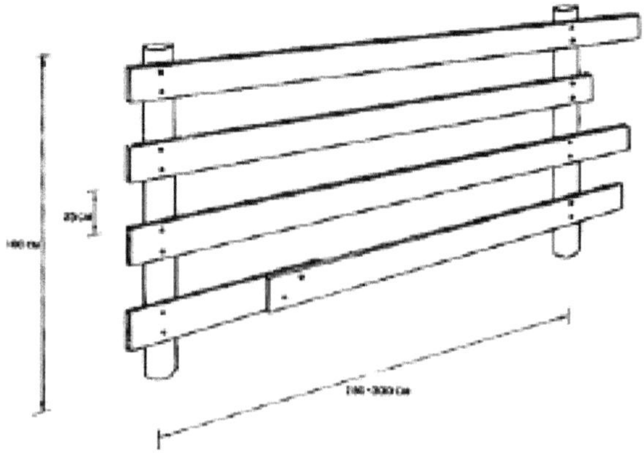

3. Permanent fencing

Lastly, *in regions with very cold winters,* or where there is no prevalent wind direction and it blows first from one corner, then from another, a temporary protection system can be envisaged involving the use of a lightweight curtain (nylon or plastic), stretching all along the shed roof to protect the manger and the ground.

Fence construction is one of the basic components of fully open yard housing. It is very limited in the combination system and virtually non-existent in the fully roofed yard system.

The permanent fencing is made up of posts, bedded in the ground, and fencing materials. Each of these can be made from different materials, but both must conform to exact building specifications.

The principal materials which can be used

For the posts, it is usually more economical to use wooden poles with a diameter of about 15 cm, but these are also less sturdy when they have to hold up under long and intensive use. *Iron posts* 15 cm in diameter and *iron pipes* 8 cm in diameter are strongest.

The actual fence can be made from a series of horizontal bars. These can either be 35 mm thick planks, or metal pipes at least 5 cm in diameter.

Despite the variety of the materials which can be used, *certain standard specifications must be observed in building the permanent fence.*

The height of the fence must be 1.60 m off the ground. This is high enough to dissuade the young cattle from trying to jump over it.

There must not be more than 25 cm of *free space between the bars*. This keeps the animals from sticking their heads through the fence, which might cause accidents and damage to the installations.

The number of bars of which the fence is made up is determined by the total height of the fence, the spacing between the bars, and the width of each bar. For 15 cm boards, 4 bars will be enough; for 50 mm metal pipes, 5 bars will be needed with the bottom bar 35 cm off the ground.

How strong the fence is depends on the *spacing of the posts*. Spacing, of course, depends on the kinds of materials used to make the fence and their dimensions. A 2.5–3 m spacing is usually considered to make the fence strong enough.

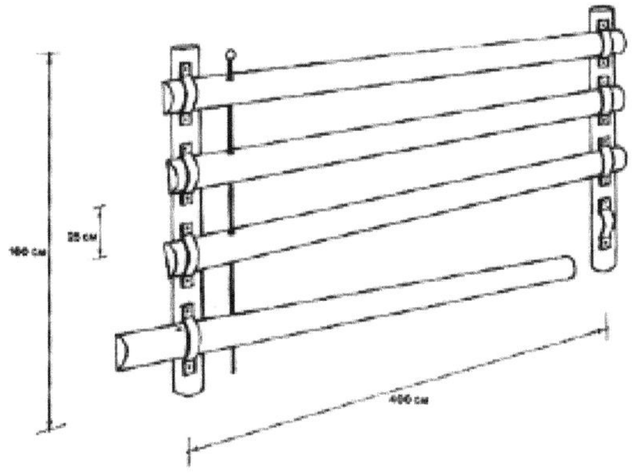

The way the fencing is joined to the posts is important. It must be strong, not risk harming the animals, and be easy to repair. The nail and barbed wire are to be avoided. The best solution is the threaded rod, and, when the fence is made entirely of wood, the bolt.

4. Movable fence

We remind the reader that all of the three suggested housing types must include a service passage so that the litter accumulating in the yard can be removed from time to time. In fully open yard housing, the service passage is directly behind the manger; in combined housing between the manger and the roofed section; in fully roofed housing beneath the roof. Except during cleaning operations, the passage is closed by movable fences.

The materials used in building the movable fence are the same as those mentioned for the permanent, or stationary fence. Height above ground: 1.60 m. Space between removable bars, 25 cm. Number of bars: 4 or 5 according to their diameter.

The only difference lies in the way the movable fence is attached to the two supporting posts. Here there are three possible solutions:

The conventional gate would not be possible, given its four-metre length, unless a lightweight, strong material such as metal pipes mounted on a reinforced axis with a support cable were used.

The rigid, movable fence is made up of horizontal bars set on vertical mounts at each end, perhaps supported by one or more slanting braces.

The fence is hinged at both ends so that it can open either way, or even be taken off altogether.

Movable bars. Whereas in permanent fencing the bars of the fence are permanently attached to the posts, with the movable fence one has to be able to take the bars away: they can slide freely through plate iron hasps attached to the posts. To make sure the bars are not moved by the animals, they can be held in place by an iron bar running through them from one end to the other.

HANDLING AND CARE OF THE ANIMALS

Loose housing is common housing, so when handling is necessary, the animals are handled as a lot. So the handling of young cattle housed in this way is quite different from that of a tied animal, especially in the case of ungelded young bulls. Loose housing therefore needs an additional annexe where the animals can be handled and held (circulation and holding passage and corral). In addition, since the animals cannot be treated on an individual basis in open yard housing, another annexe, the infirmary, should be attached for treating sick or wounded animals.

The handling and holding premises are especially essential for the periodic weighing of the lot. Without such facilities, the weighing process can become an endless, tiresome and sometimes dangerous "corrida".

So that the animals will not disperse upon exiting from the yard, they should be channelled into a circulation corridor. The principle of this corridor is a simple one. All that is needed is to build a permanent fence along the back of the housing, and parallel to it. This corridor should not be wider than 70 cm. Each yard gives on to the corridor from the entrance/exit door in the shelter wall in the rear of the yard.

The corridor ends in a corral built of permanent fencing and with a movable fence about 3 m wide to let the cattle scales through. The corral should be big enough to hold the entire lot at once, with space of about 1.50 m2 for each young bull.

When weighing starts, the movable scales is brought into the corral at the end of the holding corridor, the housing door is opened, and the animals are forced to proceed towards the scales. They can neither go back, being pushed by the ones behind, nor turn around (the 70 cm corridor is too narrow), so they have to move forward towards the scales.

After the entire lot has been weighed, all the animals are in the corral. They can be brought back into the housing area in the same way, only the scales will not be used this time.

The other lots follow the same procedure.

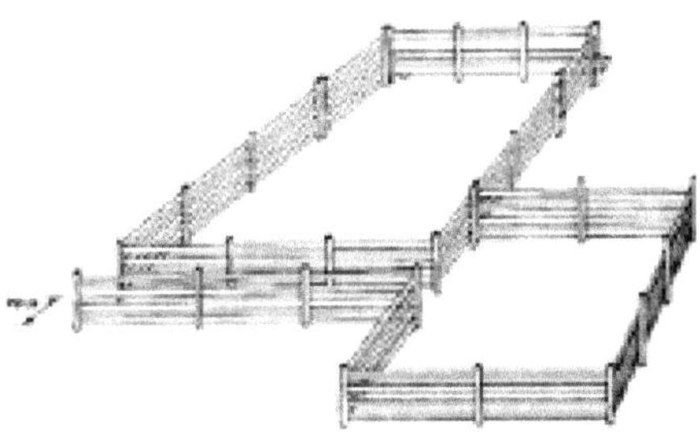

1. Handling and holding

(a) Fully open yard

The handling and holding of young cattle housed in open yards is rather exceptional, as it only occurs once a month where the animals submit to growth checks.

Partly because of this, the animals are not accustomed to be handled. If the housing does not have the proper facilities, the operation can be dangerous for both the men and the animals.

Basically, the design of these facilities must be adapted to the kind of housing, the size of the livestock unit (number of yards) and hence the number of animals. The size and weight of the animals will also be considered.

For small, open-yard livestock involving some tens of animals, the aforementioned system of adding a circulation passage 70 cm wide along the permanent fence opposite the manger side of the building is perfectly suitable. The corral is reduced to a simple enclosure of 1.5 m2/head.

For larger units, with accomodations for several hundred head, more complete installations should be planned. First of all, out of necessity housing of this type involves batteries of yards built in tandem back-to-back. It might be a good idea to run a service passage 3 m wide between them for vehicles and animals. If this is done, the best thing is to set up a corral at one end of the service passage. It should include at least three sections: *a waiting area* (1.5 m2/head) where the animals wait to be weighed; *a holding passage* 70 cm wide and about 10 m long, where the animals can be contained and which leads to the scales; *a common area* where the animals can wait after weighing (1.5 m2/head).

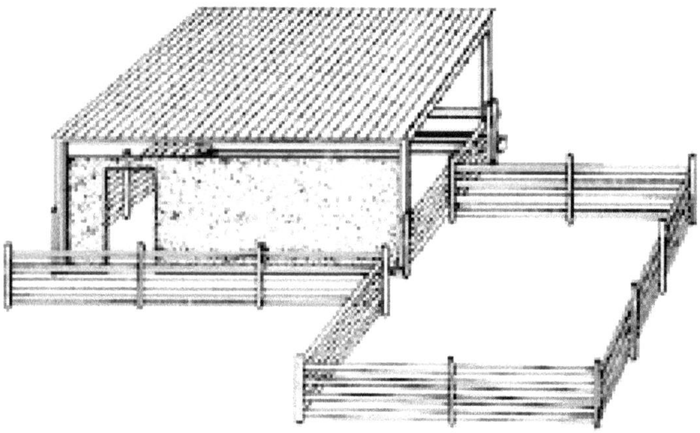

(b) Combination and fully roofed systems

For small combination and fully roofed systems holding a few tens of cattle, the system just described can be quite suitable. It boils down to a circulation passage 70 cm wide, narrow enough to double as a holding passage, and which gives onto a simple corral.

In this case, the circulation passage is always at the back of the building. The shelter wall is paralleled by a permanent fence. The animals enter the passage by the entrance/exit door already described (p.55).

For larger units with accomodations for several hundred head, facilities are of course a bit more complex. A circulation passage for each row of housing and behind the shelter walls will be retained, but there has to be a passage where all these passages lead or meet. This will be a communicating passage at one end of the battery housing, made up partially of movable fences, and giving onto a corral and single holding passage.

There really are a large number of design possibilities with regard to corrals and passages. They depend on the kinds of buildings, the number of animals to be handled, and the different operations to be carried out.

Independently of the components just described - the waiting, holding and corral areas - the complex may also include the following components:

- *a funnel area* as part of the waiting section, making it easier to get the animals into the holding passage;
- *a holding or working cage* to hold the animals still during such painful operations as foot treatments, tagging, polling, and the like;
- *a dip or shower passage* for occasional treatment of external parasites;
- *a permanent scales*; sturdier construction than the mobile models. It would be best to protect this scales by some lightweight construction;
- *a sorting arrangement* with a multiple access door to facilitate selection of individual animals and constitution of lots;
- *a loading platform* where the cattle can be loaded and unloaded onto trucks.

There is extensive literature, primarily American, detailing the various kinds of solutions.

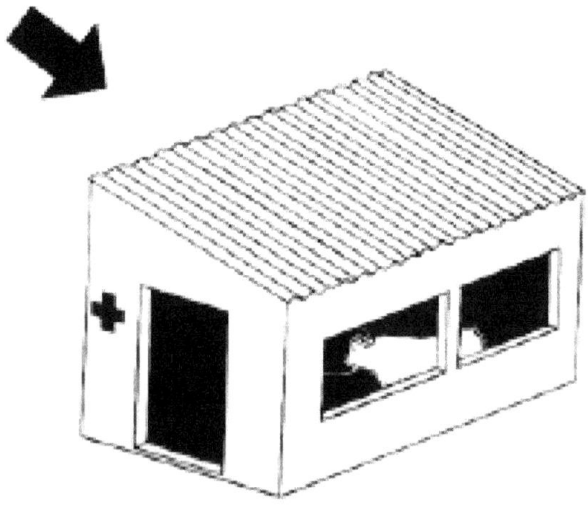

2. The infirmary

Concentrations of large numbers of young cattle in the same livestock unit necessarily entails a specific disease situation. To treat the diseases which arise, facilities must be planned as an annexe to open-yard housing, though this is not necessarily the case with stanchion stable housing.

As soon as the first symptoms appear (the animal refuses to eat, stays off by himself, is dominated by the more aggressive of his mates), he must be isolated from the rest in his lot and installed in the infirmary. If a sick animal is left with the lot in loose housing, his recovery may be jeopardized, since:

- he cannot get the proper supervision;
- he cannot get the needed treatment (veterinary care, special diet);
- the healthy animals will keep him from getting his share of the feed.

The infirmary must meet the following specifications:

- it must be so laid out that each sick animal can be isolated, either tied or in his own box where he can be untied;

- it must be so built as to take into account the fact that a sick animal needs better protection from heat, cold, wind and dampness than does a healthy one. So, preferably, the infirmary will be fully roofed even for fully open-yard or combination housing;
- in addition, this will have to be "permanent" construction (something sturdy) - concrete flooring, breeze-block walls, and the like - so that the premises can be periodically disinfected;
- it should be located near the quarters of the workers or herdsmen.

Certain afflictions require permanent surveillance and this may well occur outside of working hours;

- a certain minimum of facilities is required, such as heat, lighting, a store of veterinary medicines, individual drinking-bowl…;
- it must have accomodations for about 3 young cattle out of 100, on the average.

Building Methods

The loose housing models described in this work range from actual construction to simply working the ground. This fact may appear to justify a great deal of variety with regard to installing these facilities. Nonetheless, the actual construction has to follow a strict sequence of operations once the following preliminary work has been completed:

Site selection, which must, we remind the reader, take into account the drainage capacity of the soil as well as the proximity of fodder crops for the cattle (for silage, a distance of more than 3 km between the plot and the silos/housing complex, is definitely not recommended); the availability of water; and the existence of an all-seasons road for shipping farm products and animals off the farm.

Selection of housing system and number of units; this assumes, we remind the reader again, a detailed study of the specific conditions of the kind of livestock specialization involved.

Transport onto the site of all the building materials needed before beginning construction. Too often housing construction is halted at the last moment because certain materials are no longer available.

So, a strict sequence of building operations has to be respected, each separate job and special worker following one another in the site so that each operation is carried out under the best possible circumstances.

If this sequence is not respected, much time can be wasted. Sometimes, work already completed has to be undone or redone.

In any case the work plan and programme chosen must absolutely be respected. Any sudden modification which might disturb the smooth running of the work-site and, ultimately, satisfactory utilization of the housing, is to be shunned.

1. Site preparation

The choice of the site for building loose housing for young cattle is extremely important. The length of amortization of this kind of installation can run about ten years. It is recommended that both the present situation and development prospects for the individual farm specialization be taken into account. Particularly:

- The existence of access roads allowing all-seasons transport of farm-produced animals and feed off the farm;
- Soil type, which is, we remind the reader, very important in fully open and combination type loose housing;
- Proximity of other farm buildings, particularly those for housing the workers in charge of caring for the animals and from which operations will be directed;
- The nature of the feeding/watering system, feed storage and water supply. In effect, feed must of necessity be stored in close proximity to the housing facilities. The daily transport of fodder and water over a distance of several kilometres, often by tractor, adds very heavily onto the cost of breeder/feeder operations.

In calculating the total area of the installations, both the housing proper and the annexes must be considered: feeding passages, circulation passages and open areas facilitating the circulation and manoeuvering of transport equipement.

Next, a topographical survey including an assessment of the drainage capacities of the soil. As we have seen, this is one important essential for fully open-yard housing. The topographical survey is also essential where the land chosen for the site has to be levelled before construction.

Lastly, site preparation includes staking-out, which consists of marking off all areas of the building which touch the ground, and which include the supports for the mangers and drinking-troughs, fence posts, roof supports, and the like.

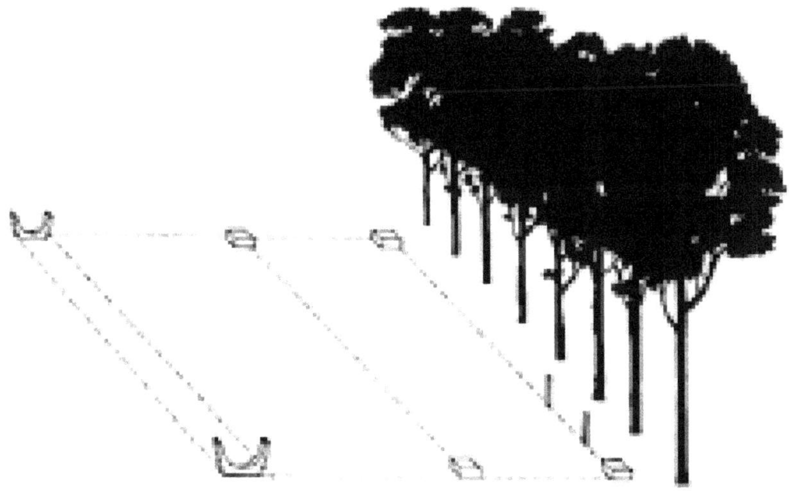

2. Installation of supports

The first step in building the housing is to install the support posts for the manger, drinking-trough and roof (if the housing is to be partly or fully roofed).

This is a two-stage process:

The *foundations can be dug* by hand, with a tractor-driven drill, or by excavator. This is also when the holes for the fence posts (both movable and permanent) should be dug.

Foundation depth for *manger supports* and *watering-trough supports* has to be calculated so as to allow these building elements to be set at the proper height.

So as to ensure that the permanent and movable fences will be strong enough, it is recommended that their support posts be sunk to a depth of about S! the portion aboveground, or roughly 55 cm.

The volume and depth of the *concrete base to support the roof* (where there is one) can vary according to the weight of the roof, and hence of the roofing materials used. With a metal frame with a bearing-surface of about 8 metres and fiber cement roof, the posts can have a volume of roughly one/eighth of a cubic meter and a depth of 50 cm.

A cement-rich concrete is needed for the *bedding of the support posts* so they will be able to bear the weight of the semi-circular pipes, which is considerable and which is increased by the weight of the water it contains in the case of the reservoir-drinking troughs.

Spacing between the *manger supports* must, we remind the reader, be calculated in line with the length of the pipes used to build the manger.

Care will be taken with the *drinking-trough supports* to ensure that the two posts which keep the pipe in a horizontal position (and hence govern its real capacity) are properly placed.

The *roof supports* usually have a metal bolt to which the frame is bolted.

3. Laying the pipes: building the roof

Let us recall that different possibilities have been considered concerning the materials which can be utilized for mangers and watering-troughs pps. 68–73). Just where *installation of the manger* fits in the building sequence has to do with the materials chosen for its construction.

Wooden or iron mangers will be installed at the end of the sequence, after the fences have been installed.

Concrete or parpen mangers, do not require prior bedding of supports, and so they can be built at this stage.

For mangers and drinking-troughs made from semi-circular irrigation pipes: they cannot possibly be laid manually, given the fact that an irrigation pipe 80 cm in diameter and 80 cm long weighs 2 tons. If unloading and laying of such pipes were attempted without the help of a crane, either the supports or the pipes themselves might very well crack.

For this reason, on the day the pipes are laid and as soon as the support base is perfectly dry, a hydraulic crane just like the ones used to load the pipes onto delivery trucks at the factory must be brought onto the farm. By planning ahead in this fashion, all the pipes to be used in the housing complex. (troughs and mangers), can be laid at the same time.

Building the roof, should there be one (fully roofed and combination systems). *At this stage the frame is erected and the roof put on.* For instance, where metal frames are used, quite a bit of room is needed to install them. Were this left until a later stage, it might be quite difficult for the workers to maneuver among the posts and fences. These should be mounted later.

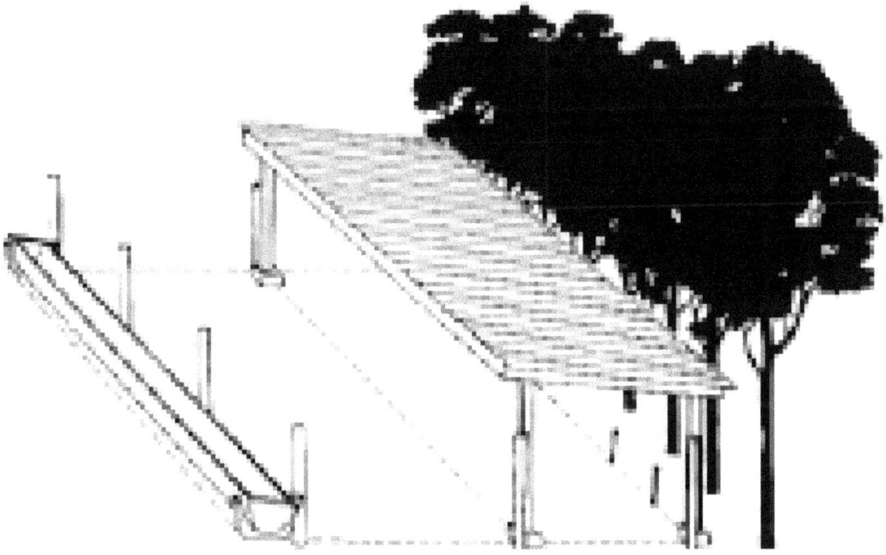

4. Bedding the fence posts

We have already indicated that the foundations for the support posts for the withers bar along the manger and reservoir-trough and for the permanent and movable fences around the housing must be dug at the very onset of the work. In fact, once the pipes have been laid and the frame built, such earthworks could only be carried out under poor conditions.

On the other hand, were the posts to be set too early, they would disturb the progress of the work and might well get broken, especially when the troughs and mangers were laid.

Not only this, but since the recommendation is to bed the manger posts so as to lay right next to the upper edge of the manger, bedding could not be carried out satisfactorily until after the pipes were laid.

Special attention must be paid to bedding the posts. The overall strength of the rent of the housing superstructure depends on how solidly these posts are laid in.

Bedding will therefore be reinforced (volume of concrete, amount of cement used in the concrete) for those posts subjected to special stress by the animals.

Basically, these *posts* are intended to support the withers bars of the manger which the animals always rush against as soon as their favourite feed, concentrates, is distributed. The same is true for the posts to support the movable fences, especially the ones barring the service passage temporarily. Not only is this an unusually wide span (4 meters instead of 2.5–3), the fact that the movable fence of course has to be hinged instead of bolted to the post makes the post particularly vulnerable.

Lastly, the posts of the holding and circulation passages also have to be bedded very solidly.

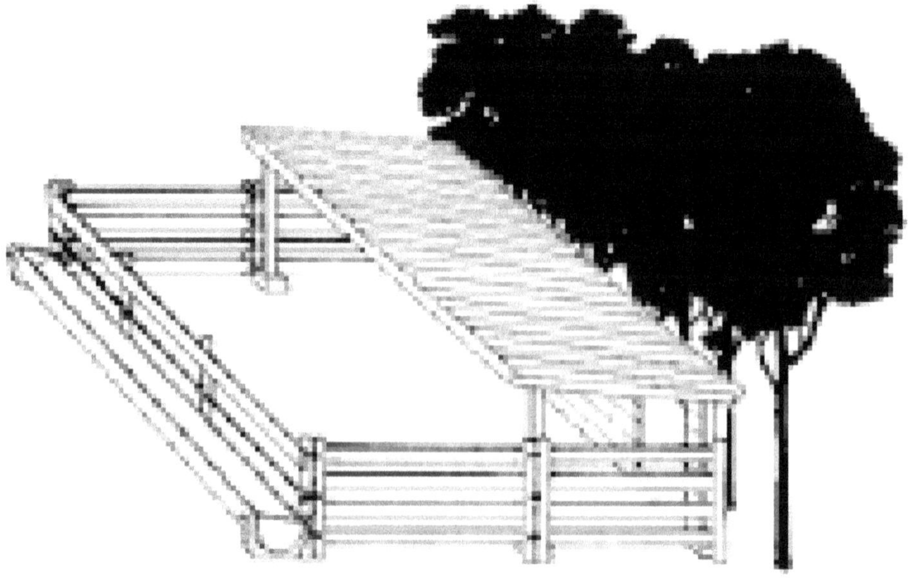

5. Floor covering and fencing

Only when all the building elements which have to be set into the ground have been installed (manger supports, roof frame, fence posts) can the *flooring work* be initiated: paving or asphalting the feeding passage, blocking or cementing the outdoor area of combination housing - as needed.

It should be noted that whenever the water for the drinking-bowls can be piped in from a network of running water, the pipes should be sunk beforehand so that they cannot be damaged by animals or machinery.

Obviously, were the flooring laid before this stage, it might be necessary to futilely destroy part of the work done in raising the supports or bedding the fence posts.

On the other hand, one thing which must absolutely be avoided is to lay the floor *after* the permanent fences and shelter wall have been built. They would so hinder the progress of the work (bringing in materials and and circulation of machinery) as to have to be temporarily taken down.

Raising the roof is the very last stage in the process, together with the installation of the withers and protection bars before the manger, the movable fences at either end of the service passage and permanent fences as planned.

At the same time, the shelter wall partition and the entrance/exit door at the covered end of roofed or semi-roofed housing will be erected.

It happens too frequently that the building of the circulation passage, upon which satisfactory handling of the animals depends, is left until some later stage. This is a mistake which must be avoided. It is really inconceivable to think of receiving the young cattle before both the housing and its annexes have been completely finished and the resting area strewn with a few bales of hay, ready for occupancy.

BUILDING BATTERY HOUSING

The housing models suggested in these pages have been put forth as self-contained units. However, it is somewhat rare that a single unit alone is built. The very modularity of the design lends itself in theory to unlimited juxtaposition, according to the production capacity of the individual farm.

There are numerous advantages to building several loose housing units in the form of one or more *battery housing* complexes. However, such construction has to be based on a certain number of principles intended to favour the input/output balance.

The advantages:

- With battery housing, several lots can be raised at the same time, which makes possible the constitution of more homogeneous lots;
- With the utilization of "continuous use" installations, wherein lots at different stages of the breeder or feeder cycle are run at the same time; revenue from the livestock operation can be regularized year-round (assuming compatability of market requirements and staggered cycles);

- With the raising of a great many animals, a scale economy over the fixed expenditure of the farm becomes possible.

  The principles:

- Absolute respect of the three types of circuits essential to the proper utilization of such housing.
- Utilization of the annexes (feeding passage, circulation passage, corral, infirmary, and so forth) for as many lots and animals as possible.
- Attempt to find solutions cutting down to the bare minimum daily distances having to be traversed by the workers and the feed transport apparatus.
- Facilitate permanent visual supervision (mangers/drinking-troughs/animals).

3

Animal Feed Management

Nationally, as well as internationally, there is a wide supply of commercial ration formulation programmes for almost any animal species. Most often the programmes are able to suggest a useable feed ration in a very short time and without to demand very much from the user besides the selection of relevant feeds from a table. For application in practise many of these programmes are appropriate, because in that case the result is what counts whereas the principles and methods behind are of no interest. Therefore commercial programmes focus on the result.

In teaching at advanced level, however, the situation is exactly the opposite. In that case, the final result is less important since the ration is not actually fed to animals anyway. Instead the basic principles of feeding and the method of combining feeds into a ration are the central element. A programme to be used in teaching should therefore put focus on the problem to be solved instead of the solution to the problem. This observation was the original starting point of the construction of a specialised programme for educational purposes.

ANIMAL FEED was originally planned to satisfy the following demands:

- Generality. The programme must be able to "speak" any language and apply to any animal species. When students have become used to the programme in for instance a course on a dairy cattle, they can apply the same programme later on in a course on pigs. Furthermore, the option of alternative languages makes the programme available to foreign students too.
- Flexibility. Decisions concerning grouping of animals, nutritional requirements, production functions, relevant feeds and "optimisation" method must be under the direct control of the user.
- Accessibility. It must be possible to use the programme without a comprehensive printed "User's Guide". On-line help must be available to all screens. Selection from menus indstead of entering of meaningless codes. In other words, the programme itself should be as anonymous as possible, so that the attention of the student is directed towards the problem to be solved, and not to the application of some computer programme.
- Shareability. The programme must be able to run in networks. Common facilities like on-line help texts or a standard feed table must be simultaneously available to all users on the network, whereas private data like grouping of animals, nutritional requirements and resulting rations must be saved strictly separate from the private data of other users.

As mentioned, a programme to be used in education should be designed in such a way that the student will focus on the problem instead of the programme. Thus all decisions must be made by the student, including tasks as definition of groups to be fed identically, definition of nutritional requirements of each group and selection of relevant feeds. The programme itself should be as anonymous as possible. The programme made, has been designed according to these guidelines. The student may determine a least-cost ration by linear programming in order to meet the requirements set up by himself. Afterwards the least-cost ration may be inspected on the screen and further edited. While the ration is being edited the nutritional contents may at any time graphically be compared to the student defined requirements in a pop-up window. If, furthermore, a production function is defined, the consequences of ration changes on production may be studied. Screen specific on-line help is available at any time.

Some animals, like cattle and sheep, are what animal scientists call "ruminants," which means that they digest their food by fermenting it in a four-chambered stomach. Inside the stomach, there are bacteria that work to break down the tough stalks of the plants the ruminant animal has eaten. This allows them to eat and digest food that other animals, including humans, cannot. The bacteria can also help produce nutrients for cattle just by living in the animal's stomach.

Ruminant animals should be fed a diet that consists mainly of forage (like grass) or roughage (like hay). When these feeds are broken down in the stomach, they can provide energy and protein. This is the diet that is most natural for the animal.

At certain times in their lives, ruminant animals may need more energy to help them produce milk or gain weight. Energy in animal feed usually comes from grains, which can be fed alone or as part of a total mixed ration (TMR). A TMR is any animal ration that has multiple kinds of feed mixed together, which prevents the animal from only eating what it likes.

When beef cattle and lambs are being prepared for market, they are sometimes fed TMRs that contain a mixture of roughage and grain. The grain in the TMR provides energy in a more concentrated form, which allows the animal to grow faster and be more efficient with its food.

Dairy cattle are also fed TMRs. These consist mostly of hay or corn silage, which is produced by grinding the whole plant at harvest and then fermenting the mixture in a silo. The grains in the TMR help the dairy cow to produce milk that is high in butterfat, which makes the milk more valuable.

Farmers need to be careful, though, because feeding too much grain and not enough forage and roughage can make a ruminant animal sick.

Non-ruminants, which are animals like chickens and pigs, digest their food more like people do. In fact, they can live without any hay or grass. Instead, they are fed mixed rations of energy and protein, which usually come from corn (energy) and soybean meal (protein).

Alternative Feeds

Feeding animals is the most expensive part of animal agriculture. To keep costs low, some farmers feed by-products from other industries to reduce the amount of hay and grain they need to buy. These products, like distiller's grains and soybean hulls, contain valuable nutrients. Feeding them to animals prevents waste and helps the farmer to save money.

Feed Additives

Feed additives are products used in animal nutrition for purposes of improving the quality of feed and the quality of food from animal origin, or to improve the animals' performance and health, e.g. providing enhanced digestibility of the feed materials. Feed additives may not be put on the market unless authorisation has been given following a scientific evaluation demonstrating that the additive has no harmful effects, on human and animal health and on the environment.

In addition to protein and energy, animals also need vitamins and minerals. Farmers can buy vitamin and mineral mixes that can be added to their animals' daily rations, or they can feed it to the animals in a solid block that the animals can lick while they are out grazing.

Calcium and phosphorous are two very important minerals in an animal's diet. Under normal circumstances, the ratio of calcium to phosphorous should be about 2:1 in order to keep the animal's body functioning properly. One way to keep this balance is to feed dicalcium phosphate, a compound which has the right balance of calcium and phosphorous for an animal diet.

Sometimes, the farmer will want to give his animals medication to treat or prevent disease. This medication can also be mixed into the animals' rations. The farmer has to be careful when he or she feeds this, because animals going to market must be taken off medicated feed weeks in advance in order to give the medicine time to wash out of their bodies. This period of time when the animals cannot have medication in their feed is called the withdrawal period.

Animal Feed

Feed is also called animal feed which is defined as food grown or developed for livestock and poultry. Modern feeds are produced by carefully selecting and blending ingredients to provide highly nutritional diets that both maintain the health of the animals and increase the quality of such end products as meat, milk, or eggs. Ongoing improvements in animal diets have resulted from research, experimentation, and chemical analysis by agricultural scientists.

Animal feed is food given to domestic animals in the course of animal husbandry. There are two basic types, fodder and forage. Used alone, the word "feed" more often refers to fodder. Feeding is the process by which organisms, typically animals, obtain food.

The evolution of different feeding strategies is varied with some feeding strategies evolving several times in independent lineages. In terrestrial vertebrates, the earliest forms were large amphibious piscivores 400 million years ago. While amphibians continued to feed on fish and later insects, reptiles began exploring two new food types, other tetrapods (carnivory), and later, plants (herbivory). Carnivory was a natural transition from insectivory for medium and large tetrapods, requiring minimal adaptation (in contrast, a complex set of adaptations was necessary for feeding on highly fibrous plant materials). History does not record when dried roughage or other stored feeds were first given to animals. Most early records refer to nomadic peoples who, with their herds and flocks, followed the natural feed supplies. When animals were domesticated and used for work in crop production, some of the residues were doubtless fed to them.

The first scientific effort to evaluate feeds for animals on a comparative basis was probably made in 1809 by the German agriculturist Albrecht von Thaer, who developed "hay values" as measures of the nutritive value of feeds. Tables of the value of feeds and of the requirements of animals in Germany followed and were later used in other countries.

Preservation of green forages such as beet leaves and corn (maize) plants by packing them in pits in the earth has long been practiced in northern Europe. The idea of making silage as a means of preserving and utilising more of the corn plant was gradually developed in Europe and was taken from France to the United States in the 1870s. When the mature, dried corn plant was fed to cattle in the winter, much of the coarse stem was wasted, but when it was chopped and ensiled (made into silage), everything was eaten. During the 20th century, concrete bunker silos for storage of silage became a common sight in many rural areas worldwide.

Raw Feeding

Raw feeding is the practice of feeding domestic dogs and cats a diet consisting primarily of uncooked meat and bones. Supporters of raw feeding believe the natural diet of an animal in the wild is its most ideal diet and try to mimic a similar diet for their domestic companions. They are commonly opposed to commercial pet foods, which they consider poor substitutes for raw feed. Opponents believe the risk of food-borne illnesses posed by the handling and feeding of raw meats would outweigh the purported benefits, and no scientific studies have been done to support the numerous beneficial claims.

Pet Food

Pet food is plant or animal material intended for consumption by pets. Typically sold in pet stores and supermarkets, it is usually specific to the type of animal, such as dog food or cat food. Most meat used for non-human animals is a byproduct of the human food industry, and is not regarded as "human grade".

Four companies-Procter & Gamble, Nestlé, Mars, and Colgate-Palmolive-are thought to control 80% of the world's pet-food market, which in 2007 amounted to US$ 45.12 billion for cats and dogs alone.

Cat and dog food sales exceeded £1.4 billion in the United Kingdom and US$14.3 billion in the United States in 2005. In 2006, the value of the UK pet food market rose to just over £1.6 billion. The leading exporters of pet food for 2004 were France ($993 million), United States ($786 million) and the Netherlands ($511 million), while the leading importers were Japan ($718 million), Germany ($617 million) and the UK ($563 million).

Fish Food

Fish foods normally contain macronutrients, trace elements and vitamins which are necessary to keep captive fish in good health. Approximately 80% of fishkeepinghobbyists feed their fish exclusively prepared foods that most commonly are produced in flake, pellet or tablet form. Pelleted forms, some of which sink rapidly, are often used for larger fish or bottom-feeding species such as loaches or catfish. Some fish foods also contain additives, such asbeta carotene or sex hormones, to artificially enhance the colour of ornamental fish.

Bird Food

Bird food is food (often varieties of seeds) eaten by birds. The most important use of bird feed globally, is as feed for domesticated poultry. However (especially in developed countries), people also make or buy bird food to feed to pet birds or use in birdfeeders for wild birds. Birdfood can be natural or commercial. The choice of what to use as birdfood depends on the species of bird being fed. Bird foods are used both in birdfeeders and to feed pet birds. It typically consist of a variety of seeds. Not all birds eat seeds. Suet as beef or mutton fat is recommended for insect-eating birds such as nuthatches andwoodpeckers. Nectar essentially sugar water attracts hummingbirds.

Poultry Feed

Poultry feed is food for farm poultry, including chickens, ducks, geese and other domestic birds. Feed for poultry mostly consists of grain. A portion of commercial feed, typically around a quarter which is known as bulkand is indigestible. The amount of bulk is referred to as bulk density. The quantity of feed, and the nutritional requirements of the feed mainly depend on the weight and the age of the poultry as well as the season. Healthy poultry require a sufficient amount of protein and carbohydrates, along with the necessary vitamins, dietary minerals, and an adequate supply of water. Certain diets also require the use of grit, tiny rocks such as pieces of granite, in the feed. Grit aids in digestion by grinding food as it passes through the gizzard. Grit is not needed if commercial feed is used.

The feed must remain clean and dry; contaminated feed can infect poultry. Damp feed encourages fungal growth. Mycotoxin poisoning, as an example, is "one of the most common and certainly most under-reported causes of toxicoses in poultry". Diseases can be avoided with proper maintenance of the feed and feeder. A feeder is the device that supplies the feed to the poultry. For privately raised chickens, or chickens as pets, feed can be delivered through jar, trough or tube feeders. The use of poultry feed can also be supplemented with food found through forageing. In industrial agriculture, machinery is used to automate the feeding process, reducing the cost and increasing the scale of farming. For commercial poultry farming, feed serves as the largest cost of the operation.

The term chicken feed is also used as a colloquialism for a small or insufficient amount, especially a small amount of money. The term was first coined and used in that sense in 1904.

Cattle Feeding

Different cattle feeding production systems have separate advantages and disadvantages. Most cows have a diet that is composed of at least someforage as grass, legumes, or silage. In fact, most beef cattle are raised on pasture from birth in the spring until autumn (7 to 9 months). Then for pasture-fed animals, grass is the forage that composes all or at least the great majority of their diet. Cattle fattened in feedlots are fed small amounts of hay or straw supplemented with grain, soy and other ingredients in order to increase the energy density of the diet. The debate is whether cattle should be raised on diets primarily composed of pasture (grass) or a concentrated diet of grain, soy, corn and other supplements. The issue is often complicated by the political interests and confusion between labels such as "free range", "organic", or "natural". Cattle raised on a primarily forage diet are termed grass-fed or pasture-raised; for example meat or milk may be called grass-fed beef or pasture-raised dairy. However, the term "pasture-raised" can lead to confusion with the term "free range", which does not describe exactly what the animals eat.

Aside from giving them the proper type of food, giving food supplements to cattle is very important especially if the forages that are giving to cattle are quite low in nutrients. Use supplements that are high in protein. Bran feeds may also use because they have high levels of nutrition.

Also, consider the reputation of feed supplier. Ask some of their present and previous customers for feedback regarding the service and the quality of the feeds that they are offering to their customer.

Dog Food

Dog food refers to food specifically intended for consumption by dogs. Technically carnivorous, dogs have sharp, pointy teeth, and have short gastrointestinal tracts better suited for the consumption of meat. In spite of this natural carnivorous design, dogs have still managed to evolve over thousands of years to survive on the meat and non-meat scraps and leftovers of human existence and thrive on a variety of foods.

Recommendations differ on what diet is best for dogs. Some people argue dogs have thrived on leftovers and scraps from their human owners for thousands of years, and commercial dog foods which have only been available for the past century contain poor-quality meats, additives, and other ingredients dogs should not ingest, or that commercial dog food is not nutritionally sufficient for their dogs.

Most store-bought pet food comes in either dry form which is also known as kibble, or wet, canned form.

In the United States alone, dog owners spent over $8.5 billion on commercially manufactured dog food in 2007. Some people make their own dog food, feed their dogs meals made from ingredients purchased in grocery or health-food stores or give their dogs a raw food diet.

Cat Food

Cats are obligate carnivores, though most commercial cat food contains both animal and plant material supplemented with vitamins, minerals and other nutrients. Cat food is formulated to address the specific nutritional requirements of cats, in particular containing the amino acid taurine, as cats cannot thrive on taurine-deficient food.

Cat food is food intended for consumption by cats. As with all species, cats have requirements for specific dietary nutrients. Certain nutrients, including many vitamins and amino acids, are degraded by the temperatures, pressures and chemical treatments used during manufacture, and hence must be added after manufacture to avoid nutritional deficiency. The amino acid taurine, for example, which is found within meat, is degraded during processing, so synthetic taurine is normally added afterwards. Long-term taurine deficiency resulting, for example, from feeding taurine-deficient dog food, may result in retinal degeneration, loss of vision, and cardiac damage.

Most store-bought cat food comes in either dry form, also known in the US as kibble, or wet canned form. Some manufacturers sell frozen raw diets and premix products to cater to owners who feed raw.

Feeding System of Carnivores

When the lion and cheetah catch a zebra they have to kill it as soon as possible.

They use their sharp claws to grip into the flesh of the zebra when they jump onto it. The zebra is then overbalanced and pulled to the ground. The big cat then puts its teeth around the zebra's windpipe and firmly closes its mouth. The zebra dies very quickly by suffocation.

Other big cats, such as the tiger, bite into the back of the neck of their prey. Their teeth move between the neck vertebrae and break the nerve cord connections to the brain. The prey dies instantly.

The teeth of these carnivores are adapted to eat they flesh of their prey. At the front of the mouth the teeth are sharp and pointed. There are four which are longer than the others. These are called the canine teeth. These are the teeth that the big cat uses to hold on to the throat of its prey to kill it.

The teeth at the back of the mouth are also sharp and pointed but larger than those at the front. When the mouth is closed these back teeth come together like the blades of a pair of scissors. These teeth are used to cut into the flesh of the prey and sheer off pieces of meat. These carnivores have powerful muscles which can open and close the mouth. Carnivores are not able to move their jaws from side to side very easily.

Carnivorous Birds

Carnivorous birds, such as the eagle, have sharp talons on their feet to take hold of their prey when they swoop down on it. They have large, curved pointed beaks with which they tear the meat from the bone.

Fishing birds can have long pointed beaks to spear fish. The stork has this type of beak. The pelican has a very strange beak. It is long and pointed like the stork's beak, but the bottom part has a bag which can expand. The pelican swims on the surface of the water with its beak open. As the bottom half of the beak moves through the water it scoops up fish which become trapped in the bag. The bag stretches as more fish are added to it.

The Frog : A carnivorous Amphibian

The frog sits very still, close to the water's edge. It is often well camouflaged and very difficult to see when it is not moving. When an insect or worm moves close enough to the frog, the amphibian quickly pushes its tongue out. The frog has a very long, sticky tongue which wraps around the prey. The frog then pulls its tongue back into its mouth. All of this happens very quickly.

The Snake : A carnivorous Retile

Snakes slither quietly and unseen towards their prey. They are well camouflaged. Many snakes have two sharp teeth called fangs which inject venom into the prey to stop it struggling. The prey is then eaten whole. The snake has no teeth with which to bite pieces from its prey or to chew it. Its jaws are very loosely held together and it can open them extremely wide. In this way it can take large frogs and other animals into its mouth and swallow them whole. Some pythons can even swallow small deer.

Carnivorous Fish

Carnivorous fish have many small, sharp teeth which they use to pull pieces of meat away from their prey. There are many rows of these teeth and they all point into the mouth. This means that when a shark bites its prey, the prey will be less likely to break free. The shark moves its body violently from side to side to help it tear meat from its prey.

The pike is a vicious carnivore and because of this it is often called the 'freshwater shark'. The pike feeds on fish, frogs and water voles. It does not eat toads because the warts on the back of toads give out a nasty tasting liquid. The pike is the longest lived of freshwater fishes. It is thought that some are as much as 70 years.

The Spider : A carnivorous Invertebrate

Most spiders are carnivores which do not actively hunt their prey. Instead, the spider sets a trap. It spins an almost invisible web which insects either fly or get blown into. When an insect is caught in the web the spider quickly injects it with venom. The venom does not kill the prey but it does stop it moving. The spider then wraps the prey in a veil of silk. If the insect was allowed to continue to struggle it could break the web or make the spider easier to see. The spider then hides and waits for another victim. It eats its fresh food later.

Human Food for Non-human Animals

Prepared foods and some raw ingredients may be toxic for animals, and care should be taken when feeding animals leftover food. It is known that the following foods are potentially unsafe for cats and dogs:

- Chocolate, coffee-based products and soft drinks
- Raisins and grapes
- Macadamia nuts
- Garlic (in large doses) and onions

Generally, cooked and marinated foods should be avoided, as well as sauces and gravies, which may contain ingredients that, although well tolerated by humans, may be toxic to animals. Xylitol, an alternative sweetener found in chewing gum and baked goods designed for diabetics, is highly toxic to cats, dogs and ferrets

ANIMAL FEED: TYPE

Animal feeds are classified as follows: (1) Concentrates, high in energy value, including fat, cereal grains and their by-products such as barley, corn, oats, rye, wheat, high-protein oil meals or cakes such as soybean, canola, cottonseed, peanut (groundnut), and by-products from processing of sugar beets, sugarcane, animals, and fish, and (2) Roughages, including pasture grasses, hays, silage, root crops, straw, and stover.

As long as livestock remains an integral part of agricultural society, there will always be the demand for animal feed on a large scale. With a large variety of livestock, from cows to chickens to horses to ducks, there must also be a large variety of different feed to accommodate their individual needs.

The main ingredients used in commercially prepared feed are the feed grains, which include corn, soybeans, sorghum, oats, and barley. Corn production was valued at nearly $25 billion in 2003, while soybean production was valued at $17.5 billion. Roughly 66 percent of sorghum production, which was valued at $965 million in 2003, is used as livestock feed. Approximately 60 percent of barley production, which totaled 227 million bushels (4,610,000 metric tons) and was valued at $765 million in 2003, is used as livestock feed. Annual oat production in 2003 was valued at $218 million.

The sale and manufacture of premixes is an industry within an industry. Premixes are composed of microingredients such as vitamins, minerals, chemical preservatives, anti-biotics, fermentation products, and other essential ingredients that are purchased from premix companies, usually in sacked form, for blending into commercial rations. Because of the availability of these products, a farmer who uses his own grain can formulate his own rations and be assured his animals are getting the recommended levels of minerals and vitamins.

The commercial sale and industrial production of animal feeds has a history dating back as far as the 1800s. It was during this time period that the significance of balanced diets and nutrients in the diets of humans and animals was beginning to make steam as a mainstream ideology. One of the first big feed producers, which was the first to manufacture corn gluten feed, was called Purina which was established in 1882. It expanded its operations near the beginning of the 20th century and still operates today. In modern times, there are three major types of feed used by farmers, each with their ingredients, uses and benefits.

Animal Feeds

Animals must receive sufficient amounts of all essential nutrients (water, energy, amino acids, vitamins, minerals) to remain healthy, to grow and to produce. Ration formulation involves combining the various ingredients so that an animal's nutritional requirements are met. Producers must always be aware that the nutritional requirements vary with species, age, production and even the season of the year. One of the major challenges facing the livestock production industry today is providing adequate amounts of a balanced ration at a reasonable cost.

Livestock feeds can be classified as roughages, cereals and supplements. Examine the individual components and mixed feeds to determine ingredients and which animal species might utilise these.

Roughages

The roughage or forage utilised in Canada comes from native rangelands and pastures, improved pastures, crops harvested and stored specifically for feeding to animals or residues of crops grown for other purposes. Roughages are usually feeds that have a low bulk density, but because of the variety of crops which are considered to be roughages, there is a great deal of variation in composition. Most feeds in this group have a high crude fibre content and a low digestible energy content.

Corn silage is one exception since, although it has high levels of crude fibre, it also has high levels of digestible energy. The protein, mineral and vitamin contents can vary substantially between roughages and also between crops of one roughage. Forages form the major part of most ruminant rations, since they are less expensive than other feeds and the ruminant animal is able to digest most plant components. However, some ruminants are fed very little roughage. Cattle and lambs being readied for market are often fed a ration based almost entirely on cereal grains. High producing dairy cows may also receive a very small roughage component in their diets.

Pasture or grazed forage is the most common roughage throughout Canada. All native and most cultivated pastures are composed of grass species alone or mixed with legumes and regional climate is the major consideration in deciding which types or combinations predominate. Temperate pastures can only be used as a feed source for a few months of the year, so forages must be harvested and stored as hay, haylage or silage to provide feed during the nongrowing seasons. Hay, haylage and green chop are all made from grasses or grass-legume mixtures.

Common legumes include alfalfa, clover and trefoil and, compared to grasses, these have higher protein and calcium levels. With hay and haylage the goal in harvesting the plants is to optimize the yield, which increases with maturity, and also to maximise digestibility, which decreases rapidly with maturity. Forage quality is very dependent on the amount of moisture in the plant at harvest which affects the nutrient loss during storage. Green chop is a forage that is cut and chopped in the field and fed fresh to confined animals so it is available only during the active growing season.

The major advantage of haylage over hay is that making hay of good quality is very weather dependent. With the advent of in-barn hay dryers, this has been somewhat alleviated but the ensiling of haylage at higher moisture levels than possible with hay is still a very popular alternative. Hays are usually stored at 15-20% moisture, while haylages may range from 45-55% moisture. The ensiling process results in a decrease in protein, and an increase in non-protein. Since haylages are usually left in the field only long enough to wilt, their exposure to the sun is much shorter, and they usually contain less Vitamin D than dry hay. They store best, with the least spoilage, in oxygen-limiting tower silos. A wide variety of compounds, classed as nutrients, preservatives, fermentation aids or biological additives have been added to haylages in an effort to improve quality after ensiling. Differing results have been obtained but it seems as though some will improve quality.

Corn silage is a popular feed in areas where corn is grown specifically for livestock feed. It is very palatable with a fairly high level of digestible energy (although this can vary somewhat depending on the amount of grain or the maturity of plants at harvest), but reasonably low levels of protein, Ca and P. Urea, anhydrous ammonia, or other sources of N can be added to increase the level of crude protein. The plants are usually harvested and put directly into the silo at approximately 60-75% moisture. Silage may also be made from the other cereal grains which usually have higher protein but lower energy content than corn so animals must eat more to gain the same amount of weight.

Corn stover and straw are examples of crop residues that may be fed to ruminants or other herbivores. Corn stover is the stalks and leaves remaining in the fields after corn is combined for grain. Waste from canneries processing sweet corn, peas, bean and other vegetable crops, may be fed effectively as ensiled crops. Most crop residues are limited in availability to a short period after the primary crop is harvested.

Cereals

Cereal grains contain substantial quantities of readily available carbohydrates, sugars, starches, fats or oils. They usually contain little fibre and are highly digestible, especially by monogastric animals, so form the major part of diets for animals with simple stomachs. Corn, wheat, oats and barley are the most common cereal grains fed in Ontario, and are generally less variable in nutrient composition than roughages. Corn has the lowest levels of protein, and is deficient in the essential amino acids lysine, tryptophan and methionine. Cereal grains are all low in Ca, and while the P level is quite high, most of it is in the form of phytic acid which is unavailable to monogastric species. Grains are stored generally at 10-15% moisture but high energy cost associated with drying has contributed to the current popularity of high moisture corn. This is stored at 22-30% moisture in oxygen-limiting, upright silos. Corn is usually infested with molds during the growing season and, under certain conditions, these organisms may produce mycotoxins that contaminate the grain and can produce disease when affected cereals are fed.

Soybeans are not generally considered cereal grains but whole beans can be fed to ruminant animals as both a protein and energy source. The protein level is 35-40% and although it is deficient in methionine it is not of much importance for ruminants. The raw beans possess a trypsin as an enzyme involved in protein degradation, inhibitor so they must be treated before they can be fed to monogastric species.

Supplements

Supplements are feeds which are high in one or more of the nutrients required by animals with the exception of water. They can be further broken down into protein supplements, energy supplements and vitamin and mineral supplements. Some feeds may fall into more than one category. Many of these feeds are by-products of industries that produce a product destined for human consumption.

Energy Supplements

Many energy-rich feeds are by-products of cereal grain milling. These by-product feeds usually come from various layers of the seed covering, which, along with being higher in fibre, are higher in protein and minerals than the seed itself, which is mainly starch. The most common milling by-products are from wheat with lesser amounts derived from corn. These are classified on their fibre content and include bran, middlings, shorts, red dog and wheat-germ meal. In Ontario and Canada, poultry feed is the largest user of these by-products. While much of the starch has been removed, they are still good sources of energy.

Molasses is the major by-product of sugar production, coming mainly from sugar cane but also from sugar beets. Most molasses in commercial use is adjusted to contain 25% water. It may also be dried so that it may be added to dry diets, but this is a much more expensive product. While molasses is a good source of trace minerals, the protein and vitamin level is quite low. It is used often to stimulate eating, to reduce dustiness in feeds, as a pellet binder, and when fortified with non-protein source, as a ruminant feed known as a Liquid Protein Supplement.

Beet pulp is another by-product of sugar extraction. It may be purchased either flaked or pelleted and may have molasses added to it. It is very palatable and, despite its relatively high fibre level, is quite digestible. It is often used in lactating cow rations to stimulate appetite.

Surplus fats may be fed as an energy source only, since they supply almost no protein, vitamins or minerals. When available at reasonable costs, fats may be used in feeds to decrease dustiness and increase palatability. Since fats are subject to oxidation and rancidity, which will decrease their palatability, antioxidants are usually added. Fats are not often fed to ruminant animals except perhaps in milk replacer formulas and occasionally in a finishing or lactating cow ration, since ruminants are not tolerant of fat levels higher than 7-8%.

Protein Supplements

These may originate from either plant or animal sources, or in the case of ruminants, from non-protein Sources. A protein supplement is a feed which has more than 20% crude protein. These are usually the most expensive constituents in animal rations so their selection is affected by cost and availability. The content and availability of amino acids for monogastric species, the presence of undesirable or toxic substances and the content of other nutrients should also be considered. Most of the plant protein-rich feeds are by-products of the oil extraction industry, with soybean meal being the most common in Ontario. The protein is usually low in the sulphur containing amino acids cysteine and methionine and may be low in lysine. Soybean meal has a protein level of 44% or 48% depending whether or not the hulls are added back. It is probably the most common protein source in North America and it is quite palatable, highly digestible and has fairly high energy levels.

Canola meal is probably the second most common oilseed meal fed in Ontario, and likely the most common in the Western provinces. It contains between 35% and 40% protein, with slightly less lysine and slightly more methionine than soybean meal. Perhaps the major disadvantages of canola meal are the high fibre levels and resulting lower energy levels for all species and the anti-nutritional factors which may be present. Canola has been bred to be low in the anti-nutritional factors that were previously associated with rapeseed, and most of the new varieties are "double-low". For most animals the best results in growth rate and feed efficiency are obtained when the level of canola meal in the diet is less than 20%.

Cottonseed meal and linseed meal are two other potential feedstuffs, but due to climate and changes in consumer preference very little is fed in Ontario. Linseed meal, often called oilcake, is a meal made from flax, which is deficient in lysine and is seldom grown in large quantities anymore. It is most popular as a feed for show cattle, since it tends to produce a very glossy hair coat, but it is generally too expensive for the protein it provides to be used commercially.

Other by-products of plant origin are brewer's and distiller's grains. Either can be purchased in the wet state, but since it is more convenient for transportation and storage, they are usually dehydrated. Distiller's dried grains contain 25-30% protein, have relatively high fibre levels and, due to the yeast present, a high B vitamin content. They are often deficient in lysine since they are primarily corn based. Brewer's dried grains have considerably more fibre (18-19%) and as a result are more suitable as a protein supplement for ruminants. The protein level is 25-30% and low in lysine, but high in methionine.

Animal protein supplements such as meat meal, blood meal, bone meal are usually by-products of the meat-packing industry. Since these are animal tissues, the amino acid distribution is usually similar to the animals' requirements, although quality may be quite variable since the composition of the meal may vary from batch to batch.

There are two major types of fishmeals including those made from the waste of fish caught for human consumption and those made from fish caught specifically for industrial use. Fish meal must be dried down to 15% moisture and have the fat content reduced to a maximum of 10%. Anti-oxidants are always added since the oil oxidizes readily. The protein is very digestible and is very high in lysine, which makes it a good protein source to supplement feed cereal grains. Fishmeal is a good source of the B vitamins, all minerals, and seems to have unidentified growth factors which enhance growth rate and feed conversion in swine.

Single cell protein. Certain bacteria, yeast, fungi and algae can ferment residues from pulp mills, food processing, alcohol production and even human or animal wastes. At the present time, harvesting and dehydration costs are uneconomical, but products resulting from this process can be fed to supply part of the protein requirement for almost all animals without any noticeable decrease in performance. The protein concentrations range from 48-80% and are reasonably well digested. Most contain moderate energy and may provide other nutrients such as the B vitamins. Perhaps as production techniques are refined, costs will decrease and these protein sources will become more competitive with other feedstuffs.

Compound Feed

One of the most common types of feeds used by commercial farmers is Compound Feed which is also known as feedstuffs, they're blends of various additives and raw materials that are formulated to specifically suit the intended animal. They're often produced as pellets or crumbles. Like modern vitamins with humans, they can be used to either satisfy the complete nutritional requirements of their target animals or as a supplement to other staples of the animals' diets. They're often complemented with extra vitamins and minerals. It is produced in astronomically large amounts, with over 600 million tons produced annually on a global scale.

Compound feeds are feedstuffs that are blended from various raw materials and additives. These blends are formulated according to the specific requirements of the target animal. They are manufactured by feed compounders as meal type, pellets or crumbles. Compound feeds can be complete feeds that provide all the daily required nutrients, concentrates that provide a part of the ration as protein as well as energy or supplements that only provide additional micronutrients, such as minerals and vitamins.

According to the American Feed Industry Association, as much as $20 billion worth of feed ingredients are purchased each year. These products range from grain mixes to orange rinds to beet pulps. The feed industry is one of the most competitive businesses in the agricultural sector, and is by far the largest purchaser of U.S. corn, feed grains, and soybean meal. Tens of thousands of farmers with feed mills on their own farms are able to compete with huge conglomerates with national distribution. Feed crops generated $23.2 billion in cash receipts on U.S. farms in 2001. At the same time, farmers spent a total of $24.5 billion on feed that year. In 2011, around 734.5 million tons of feed were produced annually around the world.

Fodder

Fodder is the another type of animal feed which is used primarily to feed domesticated livestock such as goats, sheep, cattle, horses and pigs. It is typically composed of plant matter like hay, straw and grains. The term is used to describe these plants being given to the animals after the plants have been harvested, which contrasts with forage, as will be explained below. Meat and bone meal are occasionally mixed into fodder, which has been frequently blamed for the spread of mad cow disease and has been banned in many countries.

"Fodder" refers particularly to food given to the animals including plants cut and carried to them, rather than that which they forage for themselves. It includes hay, straw, silage, compressed and pelleted feeds, oils and mixed rations, and sprouted grains and legumes.

Feed grains are the most important source of animal feed globally. The amount of grain used to produce the same unit of meat varies substantially. According to an estimate reported in 2008, "Cows and sheep need 8kg of grain for every 1kg of meat they produce, pigs about 4kg. The most efficient poultry units need a mere 1.6kg of feed to produce 1kg of chicken." Farmed fish can also be fed on grain, and use even less than poultry. The two most important feed grains are maize and soyabean, and the United States is by far the largest exporter of both, averaging about half of the global maize trade and 40% of the global soya trade in the years leading up the 2012 drought. Other feed grains include wheat, oats, barley, and rice, among many others.

Fodder is any agricultural foodstuff used specifically to feed domesticated livestock, such as cattle, goats, sheep, horses, chickens and pigs. Most animal feed is from plants, but some is of animal origin. "Fodder" refers particularly to food given to the animals including plants cut and carried to them, rather than that which they forage for themselves. It includes hay, straw, silage, compressed and pelleted feeds, oils and mixed rations, and sprouted grains and legumes.

The worldwide animal feed industry consumed 635 million tons of feed (compound feed equivalent) in 2006, with an annual growth rate of about 2%. The use of agricultural land to grow feed rather than human food can be controversial; some types of feed, such as corn (maize), can also serve as human food; those that cannot, such as grassland grass, may be grown on land that can be used for crops consumed by humans. Some agricultural byproducts fed to animals may be considered unsavory by human consumers.

- Conserved forage plants: hay and silage
- Compound feed and premixes, often called pellets, nuts or (cattle) cake
- Crop residues: stover, copra, straw, chaff, sugar beet waste
- Fish meal
- Freshly cut grass and other forage plants
- Meat and bone meal (now illegal in cattle and sheep feeds in many areas due to risk of BSE)
- Molasses
- Oligosaccharides
- Seaweed
- Seeds and grains, either whole or prepared by crushing, milling, etc.
- Sprouted grains and legumes
- Yeast extract (brewer's yeast residue)
- Native green grass
- Bran
- Concentrate mix
- Oilseed press cake (cottonseed, safflower, sunflower, soybean, peanut or groundnut)
- Green maize
- Green sorghum
- Horse gram

Traditional sources of animal feed include household food scraps and the byproducts of food processing industries such as milling and brewing. Scraps fed to pigs are called slop, and those fed to chicken are called Chicken Scratch. Brewer's spent grain is a byproduct of beer making that is widely used as animal feed.

Evolution of form

The specialisation of organisms towards specific food sources is one of the major causes of evolution of form and function, such as:

- Mouth parts and teeth, such as in whales, vampire bats, leeches, mosquitos, predatory animals such asfelines and fishes, etc.
- Distinct forms of beaks in birds, such as in hawks, woodpeckers, pelicans, hummingbirds, parrots,kingfishers, etc.
- Specialised claws and other appendages, for apprehending or killing including fingers in primates.
- Changes in body colour for facilitating camouflage, disguise, setting up traps for preys, etc.
- Changes in the digestive system, such as the system of stomachs of herbivores, commensalism andsymbiosis.

Forage

"Forage" is plant material mainly plant leaves and stems eaten by grazing livestock. Historically, the term forage has meant only plants eaten by the animals directly as pasture, crop residue, or immature cereal crops, but it is also used more loosely to include similar plants cut for fodder and carried to the animals, especially as hay or silage. Unlike fodder, which is harvested plant matter fed to animals in bunches or bales, forage is plant matter that is literally grown in a confined area with the intention of having it grased upon by various livestock. It can be composed of legumes, grasses, corn, oats, alfalfa and other edible plants. The act of eating or grazing upon the plant matter is known as Forageing.

Revenue

The feed industry is a major economical component of many countries. According to the AFIA (American Feed Industry Association), an average of $20 billion is annually spent on feed ingredients. The feed industry is also the largest purchaser of corn and soybean meal.

Filter Feeder

Filter feeders as a sub-group of suspension feeders are animals that feed by straining suspended matter and food particles from water, typically by passing the water over a specialised filtering structure. Some animals that use this method of feeding are clams, krill, sponges, baleen whales, and many fish including some sharks. Some birds, such as flamingos, are also Filter Feeders. Filter feeders can play an important role in clarifying water, and are therefore considered as ecosystem engineers.

Example- Fish

Most forage fish are filter feeders. For example, the Atlantic menhaden, a type of herring, lives on plankton caught in midwater. Adult menhaden can filter up to four gallons of water a minute; and play an important role in clarifying ocean water. They are also a natural check to the deadly red tide.

In addition to these bony fish, four shark subclass species are also filter feeders:

- The whale shark sucks in a mouthful of water, closes its mouth and expels the water through its gills. During the slight delay between closing the mouth and opening the gill flaps, plankton is trapped against the dermal denticles which line its gill plates and pharynx. This fine sieve-like apparatus is a unique modification of the gill rakers that prevents the passage of anything but fluid out through the gills, anything above 2 to 3 mm in diameter is trapped. Any material caught in the filter between the gill bars is swallowed. Whale sharks have been observed "coughing" and it is presumed that this is a method of clearing a build up of food particles in the gill rakers.
- The megamouth shark has luminous organs called that photophores around its mouth. It is believed that they may exist to lure plankton or small fish into its mouth.
- The basking shark is a passive filter feeder, filtering zooplankton, small fish and invertebrates from up to 2,000 tons of water per hour. Unlike the megamouth and whale sharks, the basking shark does not appear to actively seek its quarry, but it does possess large olfactory bulbs that may guide it in the right direction. Unlike the other large filter feeders, it relies only on the water that is pushed through the gills by swimming; the megamouth shark and whale shark can suck or pump water through their gills.
- Manta rays, also belonging to the shark subclass, can time their arrival at the spawning of large shoals of fish and feed on the free-floating eggs and sperm. This stratagem is also employed by whale sharks.

Haematophagy

Haematophagy (sometimes spelled haematophagy or hematophagia) is defined as the practice of certain animals of feeding on blood (from the Greek words, haima "blood" and phagein "to eat"). Since blood is a fluid tissue rich in nutritious proteins and lipids that can be taken without enormous effort, haematophagy has evolved as a preferred form of feeding in many small animals such as worms andarthropods. Some intestinal nematodes, such as Ancylostomids, feed on blood extracted from the capillaries of the gut and about 75% of all species of leeches e.g., Hirudo medicinalis, a free-living worm, which are haematophagous. Some fish, such as lampreys and Candirus and mammals, especially the vampire bats, and birds, such as the vampire finches, Hood Mockingbirds, andoxpeckers, also practice haematophagy.

These haematophagous animals have mouth parts and chemical agents for penetrating vascular structures in the skin of hosts, mostly of mammals, birds, and fish. This type of feeding is known as phlebotomy from the Greek words, phleps "vein" and tomos "cutting".

Once phlebotomy is performed in most insects by a specialised fine hollow "needle" called proboscis which perforates skin and capillaries; in bats by sharp incisor teeth that act as a razor to cut the skin, blood is acquired either by sucking action directly from the vases, from a pool of escaped blood, or by lapping. In order to overcome natural haemostasis (blood coagulation), vasoconstriction, inflammation, and pain sensation in the host, biochemical solutions in the saliva for instance, for pre-injection, anesthesia and capillary dilation have evolved in different haematophagous species. Anti-coagulant medicines have been developed on the basis of substances found in the saliva of several haematophagous species such as leeches.

Haematophagy can be classified into obligatory and optional practice. Obligatory haematophagous animals do not have any other type of food besides blood; one such species is Rhodnius prolixus. This contrasts with optional haematophages, like the many mosquitoes species, such as Aedes aegypti, which may also feed on pollen, fruit juice, and other biological fluids. Sometimes only the female of the species is a haematophage, this is essential for egg production and reproduction. Coyotes, wolves, and other canids may lick blood.

Haematophagy has apparently evolved independently in many disparate arthropod, annelid, nematode and mammalian taxa. For example Diptera, the insects with two wings, such as flies have eleven families with haematophagous habits more than half of the 19 haematophagous arthropod taxa. About 14,000 species of arthropods are haematophagous, even including some genera that were not previously thought to be, such as moths of the genus Calyptra. Several complementary biological adaptations for locating the hosts usually in the dark, as most haematophagous species are nocturnal and silent, in order to avoid detection and destruction by the host which have also evolved, such as special physical or chemical detectors for sweat components, CO2, heat, light, movement, etc..

Nectarivore

In zoology, a nectarivore is an animal which eats the sugar-rich nectar produced by flowering plants. Most nectarivores are insects or birds, but there are also nectarivorous mammals, notably several species of bats in the Southwestern United States and Mexico, as well as the Australian Honey Possum (T. rostratus), and Geckos (genus Phesulma) in Mauritius. The term is less exclusive than other -vores; many animals that are considered nectarivorous may also be insectivorous. Many species are nectar robbers, performing no pollination services to a plant while still consuming nectar.

Coprophagia

Coprophagia or coprophagy is the consumption of faeces, from the Greek copros, "faeces" and phagein means "to eat". Many animal species practice coprophagia as a matter of course; other species do not normally consume faeces but may do so under unusual conditions. Coprophagy refers to many kinds of faeces eating including eating faeces of other species (heterospecifics), of other individuals (allocoprophagy), or its own (autocoprophagy), those once deposited or taken directly from the anus. Coprophagous insects consume and redigest the faeces of large animals. These faeces contain substantial amounts of semi-digested food, herbivores' digestive systems are especially inefficient. The most notable faeces-eating insect is the dung-beetle and the most common is the fly.

Termites eat one another's faeces as a means of obtaining their hindgut protists. Termites and protists have asymbiotic relationship that allows the termites to digest the cellulose in their diet via the protists. It has also been proposed that hormones are passed to offspring in this way.

Pigs, like the above insects, will eat the faeces of herbivores that leave a significant amount of semi-digested matter, including their own. In certain cultures, it was common for poor families to collect horse faeces to feed their pigs. However, allowing domestic pigs to consume faeces contributes to the risk of parasite infection. The pig toilet is an ancient method of feeding pigs on garbage and human faeces, and is used in China.

Bottom Feeder

A bottom feeder is usually an aquatic animal that feeds on or near the bottom of a body of water. The body of water could be the ocean, a lake, a river, or an aquarium. Bottom feeder is a general term which is used particularly in the context of aquariums. Biologists often use the term benthos, particularly for bottom feeding invertebrates, such as shellfish, crabs, crayfish, sea anemones, starfish, snails, bristleworms and sea cucumbers. However, the term benthos includes all aquatic life that lives on or near the bottom, which means it also includes non-animals, such as plants and algae.

Biologists also use specific terms that refer to bottom feeding fish, such as demersal fish, groundfish, benthic fish and benthopelagic fish. Examples of bottom feeding fish species groups are flatfish such as halibut, flounder, plaice, sole, eels, cod, haddock, bass, grouper, carp, bream (snapper) and some species of catfish and shark.

Bottom feeders are not necessarily detritivores, although there are many that are. Some bottom feeders graze along the bottom, feeding on plant material. Other bottom feeders may feed on other bottom feeders, and thus are carnivores.

Many bottom feeders are also capable of burrying themselves. A variety of invertebrates are able to bury themselves, such as bristleworms, sea cucumbers, and snails. Many vertebrate bottom feeders may bury themselves, such as flatfish or stingray.

In many bottom feeders, a mechanism to deal with substrate is often necessary. In the case of some organisms such as sea cucumbers, the sand is usually passed through the body. In fish, sand will be pumped out of the mouth through the gill slit.

In fish, most bottom feeders exhibit a flat ventral region so as to more easily rest their body on the substrate. The exception may be the flatfish, which are laterally depressed but lie on their sides. Also, many exhibit what is termed an "inferior" mouth, which means that the mouth is pointed downwards; this is beneficial as their food is often going to be below them in the substrate. Those bottom feeders with upward-pointing mouths, such as stargazers, tend to seize swimming prey. Some flatfish such as halibut actually have a "migrating" eye that moves to the upward-facing side of the fish as it ages.

Folivore

In zoology, a folivore is a herbivore that specialises in eating leaves. Mature leaves contain a high proportion of hard-to-digest cellulose, less energy than other types of foods, and often toxic compounds. For this reason, folivorous animals tend to have long digestive tracts and slow metabolisms. Many enlist the help of symbiotic bacteria to release the nutrients in their diet. Additionally, it has been observed in folivorous primates, animals may exhibit a strong preference for immature leaves, which tend to be higher in energy and protein and lower in fibre and poisons than more mature leaves. It has been observed that folivory is extremely rare among flying animals. Morton (1978) attributed this to the fact that leaves are heavy, slow to digest, and contain little energy relative to other foods. The Hoatzin is an example of a flighted, folivorous bird.

Equine Nutrition

Equine Nutrition is defined as the feeding of horses, ponies, mules, donkeys, and other equines. Correct and balanced nutrition is a critical component of proper horse care.

Horses are non-ruminant herbivores of a type known as a "hindgut fermenter." This means that horses have only one stomach, as do humans. However, unlike humans, they also have to digest plant fibre (largely cellulose) that comes from grass and hay. Therefore, unlike ruminants, which digest fibre in plant matter by use of a multi-chambered stomach, horses use microbial fermentation in a part of the digestive system known as the caecum (or cecum) to break down the cellulose. In practical terms, horses prefer to eat small amounts of food steadily throughout the day, as they do in nature when grazing on pasture. Although this is not always possible with modern stabling practices and human schedules that favour feeding horses twice a day, it is important to remember the underlying biology of the animal when determining what to feed, how often, and in what quantities.

The digestive system of the horse is somewhat delicate. Horses are unable to regurgitate food, except from the oesophagus. Thus, if they overeat or eat something poisonous, vomiting is not an option. They also have a long, complex large intestine and a balance of beneficial microbes in their caecum that can be upset by rapid changes in feed. Because of these factors, they are very susceptible to colic, which is a leading cause of death in horses. Therefore, horses require clean, high-quality feed, provided at regular intervals, and may become ill if subjected to abrupt changes in their diets. Horses are also sensitive to molds and toxins. For this reason, they must never be fed contaminated fermentable materials such as lawn clippings. Fermented silage or "haylage" is fed to horses in some places; however, contamination or failure of the fermentation process that allows any mold or spoilage may be toxic.

Suction Feeding

Suction feeding is a method of ingesting a prey item in fluids by sucking the prey into the predator's mouth. This is typically accomplished by the predator expanding the volume of its oral cavity and/or throat, resulting in a pressure difference between the inside of the mouth and the outside environment. When the mouth is opened, the pressure difference causes water to flow into the predator's mouth, carrying the prey item in with the fluid flow. Bony fish have evolved ingenious mechanisms, so-called mechanical linkages for rapid and forceful expansion of the buccal cavity.

Ophiophagy

Ophiophagy is a specialised form of feeding or alimentary behaviour of animals which hunt and eat snakes. There are ophiophagous mammals such as the skunks and the mongooses, birds such as snake eagles, the Secretary Bird, and some hawks, lizards such as theCommon collared lizard, and even other snakes, such as the Central and South American mussuranas and the North American Common Kingsnake. The genus of the venomous King Cobra (Ophiophagus hannah) is named for this habit.

In some regions, farmers keep ophiophagous animals as pets in order to keep their living environment clear of such snakes as cobras and pit vipers including rattle snakes and lanceheads which annually claim a large number of deaths of domestic animals, such as cattle, and bites on humans. An example is tamedmongooses in India. In the 1930s, a Brazilian plan to breed and release large numbers of mussuranas for the control of pit vipers was tried but didn't work. The Butantan Institute, in São Paulo, which specialises in the production of antivenoms, erected a statue of the mussurana Clelia, clelia as its symbol and a tribute to its usefulness in combating venomous snake bites. Peafowl have been kept for millennia due to their ophiophagous habit.

Ram Feeding

Ram feeding is also known as lunge feeding. It is a method of feeding underwater in which the predator moves forward with its mouth open, engulfing the prey along with the water surrounding it. During ram feeding, the prey remains fixed in space, and the predator moves its jaws past the prey to capture it. The motion of the head may induce a bow wave in the fluid which pushes the prey away from the jaws, but this can be avoided by allowing water to flow through the jaw. This can be accomplished by means of an expandable throat, as insnapping turtles and baleen whales, or by allowing water to flow out through the gills, as in sharks andherring. A number of species have evolved narrow snouts, as in gar fish and water snakes .

Herrings often hunt copepods. If they encounter copepods schooling in high concentrations, the herrings switch to ram feeding. They swim with their mouth wide open and their opercula fully expanded. Every several feet, they close and clean their gill rakers for a few milliseconds called filter feeding. The fish all open their mouths and opercula wide at the same time, the red gills are visible in the photo above-click to enlarge. The fish swim in a grid where the distance between them is the same as the jump length of the copepods.

Pivot Feeding

Pivot feeding is a method to transport the mouth towards the prey by an upward turning of the head, which is pivoting on the neck joint. Pipefish such as sea horses and sea dragons are specialised on this feeding mechanism. With prey capture times of down to 5 ms (shrimpfish Centriscus scutatus) this method is used by the fastest feeders in the animal kingdom.

The secret of the speed of pivot feeding is in a locking mechanism, in which the hyoid arch is folded under the head and is aligned with the urohyal which connects to the shoulder girdle. The trigger mechanism of unlocking is debated, but is probably in lateral adduction.

Myrmecophagy

Myrmecophagy is a feeding behaviour defined by the consumption of termites or ants, particularly as pertaining to those animal species whose diets are largely or exclusively composed of said insect types. Literally, myrmecophagy means "ant eating" (Ancient Greek: murmekos means "ants" and phagein means "to eat") rather than "termite eating" (for which the strict term is termitophagy). However, the two habits often overlap, as both of these eusocial insect types often live in large, densely populated nests requiring similar adaptations in the animal species that exploit them. Myrmecophagy is found in a number of land-dwelling vertebrate taxa, including reptiles and amphibians such as horned lizards and blind snakes, narrow-mouthed toads of the family Microhylidae and posion frogs of the Dendrobatidae, and several mammalian groups such as anteaters, aardvarks, aardwolves, pangolins,sloth bears, and many armadillos, numbats and echidnas, for instance.

Otherwise unrelated mammals that specialise in myrmecophagy often display similar adaptations for this niche. Many have powerful forelimbs and claws adapted to excavating the nests of ant or termite colonies from the earth or from wood or under bark. Most have reduced teeth and some have reduced jaws as well. Practically all have long, sticky tongues. In the nineteenth and early twentieth century, many zoologists saw these shared features as evidence of relatedness, and accordingly they regarded the various species as single order of Mammalia, the Edentata.

Lunge Feeding

Baleen whales feed on plankton by a technique called Lunge Feeding. Lunge feeding could be regarded as a kind of inverted suction feeding, during which a whale takes a huge gulp of water, which is then filtered through the baleens. Biomechanically, this is a unique and extreme feeding method, for which the animal at first must accelerate to gain enough momentum to fold its elastic throat (buccal cavity) around the to be swallowed water volume. Subsequently, the water flows back through the baleens keeping back the food particles. The highly elastic and muscular buccal rills are a specialised adaptation to this feeding mode.

Myzocytosis

Myzocytosis (from Greek: myzein meaning "to suck" and kytos meaning "container", hence referring to "cell") is a method of feeding found in some heterotrophic organisms. It is also called "cellular vampirism" as the predatory cell pierces the cell wall and/or cell membrane of the prey cell with a feeding tube, the coinoid, sucks out the cellular content and digests it.

Myzocytosis is found in Myzozoa and also in some species of Ciliophora both comprise the alveolates. A classic example of myzocytosis is the feeding method of the infamous predatory ciliate, Didinium, where it is often depicted devouring a hapless Paramecium. The suctorian ciliates feed exclusively through myzocytosis, sucking out the cytoplasm of prey via superficially drinking straw-like pseudopodia.

Phagocytosis

In cell biology, phagocytosis is the process of engulfing a solid particle by a phagocyte or a protist to form an internal phagosome. Phagocytosis was revealed by Ilya Mechnikov in 1882. Phagocytosis is a specific form of endocytosis involving the vesicular internalisation of solids such as bacteria, and is, therefore, distinct from other forms of endocytosis such as the vesicular internalisation of various liquids. Phagocytosis is involved in the acquisition of nutrients for some cells, and, in the immune system, it is a major mechanism used to remove pathogens and cell debris. Bacteria, dead tissue cells, and small mineral particles are all examples of objects that may be phagocytosed.

The process is homologous to eating only at the level of single-celled organisms; in multicellular animals, the process has been adapted to eliminate debris and pathogens, as opposed to taking in fuel for cellular processes, except in the case of the animal Trichoplax.

Lepidophagy

Lepidophagy is a specialised feeding behaviour in fish that involves eating of scales of other fish. Lepidophagy is widespread, having been independently evolved in at least five freshwater families and seven marine families. Lepidophagy has been reported in a range of fish including: Chanda nama belongs to family Ambassidae,

Terapon jarbua belongs to family Terapontidae, several marine catfish belongs to family Ariidae, some piranha, Exodon paradoxus and Roeboidesspecies belongs to family Characidae, along with Perissodus eccentricus, Perissodus microlepis, Plecodus elaviae, Plecodus multidentatus, Plecodus paradoxus and Plecodus straeleni belongs to family Cichlidae.

Fish scales are a surprisingly nutritional food source, containing layers of keratin and enamel, as well as a dermal portion and a layer of protein-rich mucus. They are a rich source of calcium phosphate. However, the energy expended to make a strike versus the amount of scales consumed per strike puts a limit on the size of the lepidophage; such fish seldom exceed 20 centimetres (7.9 inches) and most are under 12 cm (4.7 in). There are a number of advantages to consuming scales: scales are common, covering the body of most fish species, can be regrown relatively quickly by "prey" fish, are abundant and seasonally reliable, and their removal requires specific behaviours or morphological structures. A diversity of morphologies and attack behaviours are used by lepidophagous predators. The behavioural origins of scale feeding may be different for different lineages.

Trophallaxis

Trophallaxis is the transfer of food or other fluids among members of a community through mouth-to-mouth (stomodeal) or anus-to-mouth (proctodeal) feeding. It is most highly developed in social insects such as ants, termites, wasps and bees. The word was introduced by the entomologist William Morton Wheeler in 1918. The behaviour was used in the past to support theories on the origin of sociality in insects.

In ants such as the species Solenopsis invicta, individual colony members store food in their crops and regularly exchange it with other colony members and larvae to form a sort of "communal stomach" for the colony. In termites and cockroaches, proctodeal trophallaxis is crucial for replacing the gut endosymbionts that are lost after every molt. This should not be confused with coprophagia.

Some vertebrates such as birds and wolves also feed their young through trophallaxis. Also vampire bats are noted for this behaviour. Trophallaxis serves as a means of communication, at least in bees and ants. In some species of ants, it may play a role in spreading the colony odour that identifies members.

Self-cannibalism

Self-cannibalism is defined as the practice of eating oneself which is also called auto-cannibalism, or auto-sarcophagy. A similar term which is applied differently is autophagy, which specifically denotes the normal process of self-degradation by cells. While almost an exclusive term for this process, autophagy nonetheless has occasionally made its way into more common usage. The short-tailed cricket is known to eat its own wings. There is evidence of certain animals digesting their own nervous tissue when they transition to a new phase of life. The sea squirt with a tadpole-like shape contains a ganglion 'brain' in its head, which it digests after attaching itself to a rock and becoming stationary, forming an anemone-like organism. This has been used as evidence that the purpose of brain and nervous tissue is primarily to produce movement. Self-cannibalism behaviour has been documented in North American rat snakes where one captive individual twice attempted to consume itself, dying in the second attempt. Another wild rat snake individual was found having swallowed about two-thirds of its body.

Oophagy

Oophagy, literally "egg eating", is the practice of embryos feeding on eggs produced by the ovary while still inside the mother's uterus. The word oophagy is formed from the classical Greek oion, egg and classical Greek phagein, to eat.

Oophagy is thought to occur in all sharks in the order Lamniformes and has been recorded in the bigeye thresher (Alopias superciliosus), the pelagic thresher (A. pelagicus), the shortfin mako (Isurus oxyrinchus) and the porbeagle (Lamna nasus) among others. It also occurs in the tawny nurse shark(Nebrius ferrugineus), and in the family Pseudotriakidae.

This practice may lead to larger embryos or prepare it for a predatory lifestyle. There are variations in the extent of oophagy among the different shark species. The grey nurse shark (Carcharias taurus) practices intrauterine cannibalism, the first developed embryo consuming both additional eggs and any other developing embryos. Slender smooth-hounds (Gollum attenuatus), form egg capsules which contain 30-80 ova within which only one ovum develops while all other ova are ingested and packed to an external yolk sac. The embryo then develops normally without ingesting further eggs.

Oophagy is also used to refer to more general egg-eating behaviours such as those practised by some snakes or the differential oophagy practised by the Polistes biglumis species of wasp. Oophagy has been observed in Leptothorax acervorum as well.

Ovophagy

Ovophagy is a rare though widely observed condition in the reproduction of certain animals, whereby fetuses devour each other before birth. Subsequent research has shown ovophagy to be sporadic and rare among the vast majority of animal species.

Paedophagy

Paedophagy literally meaning the "consumption of children". It is defined as the feeding behaviour of fish or other animals whose diet is partially, or primarily the eggs or larvae of other animals. Various species of cichlid fish from Lake Malawi of the genera Caprichromis, Hemitaeniochromis, and Naevochromis are paedophages. In Lake, Tanganyika species of the genus Haplotaxodon and possibly

Greenwoodochromis bellcrossi may also be implicated in this sort of feeding strategy. The orange chromide, (Etroplus maculatus) is also a paedophagous cichlid.

Some paedophagus fish specialise in feeding on the young of mouthbrooding cichlids, ramming the buccal cavity of the mouthbrooder to dislodge, and then consume the eggs or larvae held within.

Placentophagy

Placentophagy (also referred to as placentophagia) is the act ofmammals eating the placenta of their young after childbirth.

The placenta contains high levels of prostaglandin which stimulates involution as an inward curvature or penetration, or, a shrinking or return to a former size of the uterus, in effect cleaning the uterus out. The placenta also contains small amounts of oxytocin which eases birth stress and causes the smooth muscles around the mammary cells to contract and eject milk. There is no scientific evidence, however, that eating placenta provides these hormonal effects.

There is also a school of thought that holds that placentophagy naturally occurred to hide any trace of childbirth from predators in the wild. Most placental mammals participate in placentophagy, including herbivores; exceptions include Pinnipedia, Cetacea, the camel and (in most cases) the human. Placentophagy has been observed in Insectivora, Rodentia, Chiroptera, Lagomorpha, Carnivora, Perissodactyla, Artiodactyla with the camel as a noted exception, and Primates. Marsupials, which are an order of metatherian (pouched) mammals, resorb rather than deliver the placenta, and therefore cannot engage in placentophagia; they do, however, vigorously lick birth fluids as they are excreted.

Some research has shown that ingestion of the placenta can increase the pain threshold in pregnant rats. Rats that consumed the placenta experienced a modest amount of elevation of naturally-occurring opioid-mediated analgesia. Endogenous opioids, such as endorphin and dynorphin, are natural chemicals, related to the opium molecule, that are produced in the central nervous system. Production of these endogenous opioids is increased during the birthing process. They have the ability to raise the threshold of pain tolerance in the mother. When coupled with ingested placenta or amniotic fluid, the opioid effect on pain threshold is dramatically increased. Rats that were given meat instead of the placenta showed no increase in the pain threshold. There have been no scientific studies which show that placentophagy enhances analgesia in humans.

Breastfeeding

Breastfeeding is the feeding of an infant or young child with breast milk directly from female human breasts (i.e., via lactation) rather than using infant formula from a baby bottle or other container. Babies have asucking reflex that enables them to suck and swallow milk. Experts recommend that children be breastfed within one hour of birth, exclusively breastfed for the first 6 months, and then breastfed until age two with age-appropriate, nutritionally adequate and safe complementary foods. Some working mothers express milk to be used while their child is being cared for by others.

Breastfeeding was the rule in ancient times up to recent human history, and babies were carried with the mother and fed as required. With 18th and 19th century industrialisation in the Western world, mothers in many urban centers began dispensing with breastfeeding due to work requirement in urban Europe. Breastfeeding declined significantly from 1900 to 1960, due to improved sanitation, nutritional technologies, and increasingly negative social attitudes towards the practice. Under modern health care, human breast milk is considered the healthiest form of milk for babies. From the 1960s onwards, breastfeeding experienced a revival which continues to the 2000s, though some negative attitudes towards the practice still remain.

Breastfeeding promotes health for both mother and infant and helps to prevent disease. Longer breastfeeding has also been associated with better mental health through childhood and into adolescence. Experts agree that breastfeeding is beneficial and have concerns about the effects of artificial formulas. Artificial feeding is associated with more deaths from diarrhoea in infants in both developing and developed countries.

Cannibalism

In zoology, Cannibalism is defined as the act of an individual of a species consuming all or part of another individual of the same species as food. Cannibalism is a common ecological interaction in the animal kingdom and has been recorded for more than 1500 species. It does not, as once believed, occur only as a result of extreme food shortages or artificial conditions, but commonly occurs under natural conditions in a variety of species. Cannibalism seems to be especially prevalent in aquatic communities, in which up to approximately 90% of the organisms engage in cannibalism at some point of the life cycle. Cannibalism is also not restricted to carnivorous species, but is commonly found in herbivores and detritivores.

Palynivore

In zoology, Palynivore is an herbivorous animal which selectively eats the nutrient-rich pollen produced by angiosperms and gymnosperms. Most true palynivores are insects or mites. The category in its strictest application includes essentially all bees, and a few different kinds of wasps, as pollen is often the only solid food consumed by all life stages in these insects (palynivorous mites and thripstypically feed on the liquid content of the pollen grains without actually consuming the exine). The list is expanded greatly if one takes into consideration species where either the larval or adult stage feeds on pollen, but not both; there are other wasps which are in this category, as well as many beetles, flies, butterflies, and moths. There is a vast array of different insects that will feed opportunistically on pollen, as will various birds and other nectarivores.

Animal Nutrition

All living organisms require specific essential nutrients to satisfy the biological processes associated with tissue maintenance and repair, for growth and for all other productive activities including reproduction, lactation or work. Unlike green plants, animals cannot capture solar energy and combine this with basic elements to provide nutrients but must rely on finding, ingesting and digesting suitable feeds to satisfy their requirements. Most potential feeds have complex chemical structures that must be broken down (digested) into simple compounds before they can be taken into (absorbed) and used within the animal body. This process includes the ingestion of feeds, the physical and chemical reduction to simple products for absorption from the digestive tract and the subsequent elimination of indigestible residues. Chemically, digestion involves a hydrolytic reaction splitting large molecules until they are reduced to very small components that can pass across the intestinal lining into the body. Combinations of voluntary and involuntary mechanisms under both neural and endocrine controls and with acceleration by enzyme catalysts regulate these processes.

FEED DIGESTIVE SYSTEM

Livestock are macrophageous animals ingesting pieces of plant or animal material and, once inside the alimentary tract, the complex feed is broken down by digestive processes for absorption into the body. Consuming stationary plants seems a relatively simple means of satisfying feed requirements but is in fact rather difficult since most animals cannot digest the complex cell walls. Feeding on animal tissue presents substantially fewer digestive challenges. Cells forming animal tissues contain nuclei and other organelles in a fluid or semi-fluid matrix (cytoplasm) bounded by a simple plasma membrane. After ingestion by another animal, this plasma membrane breaks easily, liberating glycogen, protein and other constituents for digestion by enzymes secreted from the lining of the digestive system.

Most plant cells, in addition to their cytoplasmic matrix with its associated organelles and plasma membranes, posses an exoskeleton to provide mechanical support. Carbon, transformed into organic form through photosynthesis, may be used immediately for metabolic needs, converted into structural material or stored as complex polysaccharides. Some of the energy reserve is conserved as starch in cereal grains, roots or tubers. Most of the carbohydrate must, however, be synthesised into the supporting tissues. These cell walls are composed of a cellulose fibre network in a gel like matrix of hemicellulose and lignin, with relative proportions shifting from the former to the latter as plants mature.

Digestion includes the ingestion of feeds, the physical and chemical reduction to simple products for absorption from the digestive tract and the subsequent elimination of indigestible residues. The processes involved are highly regulated by combinations of voluntary and involuntary mechanisms under both neural and endocrine control. Plant-consuming animals obtain energy from breaking down starch and cellulose. Although these two compounds differ only in bond linkages between glucose molecules, this seemingly small variation produces substantial differences in physical properties and suitability as components for inclusion in diets for particular animal species.

Animals are usually classified into two groups on the basis of their digestive physiology; ruminants and monogastric (or non-ruminants). This is really an over-simplification, since some non-ruminants, in addition to their simple stomachs, have an important symbiotic relationship with a microbial population in their hind-gut. Animals such as the horse and the rabbit are able to take advantage of the nutrients from many of the same feeds as ruminants. The major difference between these animals and ruminants is that their fermentation takes place at the end of the gastrointestinal tract while in the ruminant it occurs at the beginning. These non-ruminant herbivores also have requirements for dietary essential amino acids and vitamins, since although they are synthesised by the microbes, the animal's ability to absorb them from the large intestine is severely limited.

Canines and felines have simple stomachs, short intestinal tracts and undeveloped caeca. All ingested food passes quickly through the digestive system with little opportunity for microbial fermentation. Thus, these animals are carnivores, feeding almost entirely on animal tissue although some nutrients can also be obtained from high quality plant material like hulled grains or tubers that contain starch rather than cellulose. Omnivores, like humans or pigs, also possess simple stomachs but their digestive systems are capable of extracting nutrients from readily digested plant tissue if the plant cell wall is thin or the fibrous coat surrounding grain seeds or nuts is disrupted or removed. Omnivores vary considerably in ability to obtain nutrients from plant cell walls. In pigs, the sacculated colon and caecum make up about half of the total digestive capacity, providing substantial volume for microbial fermentation. This allows extensive break-down of complex plant tissues and liberation of sufficient nutrients from reasonable quality forages to satisfy maintenance requirements of mature pigs. In contrast, microbial fermentation is possible in only 17% of the human tract, severely limiting ability to digest fibre.

Some herbivores like cows, sheep, goats, buffaloes and deer, posses a complex fore-stomach (rumen, reticulum and omasum) populated by micro-organisms. This is a symbiotic relationship since the bacteria and protozoa inhabiting the rumen posses the enzymes necessary to ferment and break down cellulose, liberating volatile fatty acids fore absorption directly through the rumen wall and used as an energy source by the host. The micro-organisms also synthesise nutrients into their own tissues and, as their population increases, large numbers are flushed into the true stomach (abomasum) and intestines for digestion by the animal. Alternately, complex or protected proteins, peptides and even some carbohydrates may by-pass the rumen and escape fermentation. These pass directly into the abomasum and intestines to release nutrients through conventional mammalian digestion-absorption processes. Species with this modification are ruminants and can obtain all essential nutrients from plant cellulose. Non-ruminant herbivores such as equines, cameloids and rabbits can also derive substantial nutrients from plant materials. These animals have well developed caeca and large intestines that contain numerous symbiotic micro-organisms possessing the ability to ferment cellulose.

RUMINANT DIGESTIVE SYSTEM

Ruminants, unlike most other mammals, do not have upper incisor teeth, but possess a tough dental pad on which the bottom incisors can put pressure. In cattle, the tongue is used to pull long feed into the mouth, while the lips are used when the animal is grazing or eating smaller feeds such as grains. Usually enough chewing occurs to mix the feed with saliva and form a bolus which may be swallowed. Domesticated ruminants produce large amounts of alkaline saliva, which contains a little amylase, an enzyme required for starch digestion. The saliva aids the chewing and swallowing of dry feeds, allows for breakdown of short chain triglycerides by salivary lipase, and buffers acids produced during rumen fermentation. The volume of saliva produced is affected by the form of the feed being consumed and the water intake of the animal.

The rapid movement of the bolus through the oesophagus is probably due to the presence of striated muscle in the ruminant oesophagus. Three sphincters are also present but they may be more important during regurgitation than in swallowing. The bolus is deposited into the anterior portion of the rumen or the area of the reticulo-rumen fold.

The ruminant stomach is divided into four compartments, the reticulum, rumen, omasum and abomasum. The reticulum and rumen are joined by a fold of tissue; the reticulo-rumen fold, with the result that digesta may move freely between the two compartments. The ruminant stomach contains a large microbial population that exists in a symbiotic (or mutually beneficial) relationship with the animal. The walls of the rumen are lined with epithelial cells which terminate in papillae (small, finger-like projections). These increase the surface area for absorption and for microbial contact with the ingesta. The rumen microflora ferment cellulose, liberating volatile fatty acids. These volatile fatty acids (mainly propionic, butyric and acetic acid) are absorbed directly through the wall and serve as the main energy source of ruminants.

Rumination, or cud chewing (the ability to do so being one of the characteristics of ruminants), is the regurgitation, chewing to reduce particle size and swallowing of rumen contents. More time is spent chewing and more saliva is secreted during rumination than when the animal is eating. Animals ruminate much more during the night than they do during the daylight. Contraction of the muscular pillars in the rumen and the reticulum produce rhythmic mixing and moving of the ingesta throughout the rumen and reticulum. In young ruminants the reticular or esophageal groove is a continuation of the esophagus which forms a tube to bypass the reticulo rumen. This allows milk to pass directly into the omasum and then to the abomasum where it may be easily digested.

The next compartment is the omasum whose interior is partially filled by a number of longitudinal folds or leaves. The spaces between the leaves are packed tightly with finely ground ingesta, giving the whole compartment a round, hard appearance. The function of the omasum has not been clearly defined, but it is thought that it may act as a filter through which water and fine particles may enter the abomasum, but coarse particles may not. It also absorbs water and some ions from the digesta.

The last stomach compartment is the abomasum, which is very similar in structure and function to the glandular stomach in non-ruminants. Acid and some digestive enzymes are added to the stomach contents as they are moved through by contractions of the muscular wall.

The acidic ingesta passes into the duodenum where it is mixed with bile and pancreatic secretions. There is little digestion of soluble carbohydrates and lipids since most have been destroyed in the rumen. The small intestine is very efficient in absorbing amino acids. The large intestine and caecum may be of lesser importance in the ruminant animal, since many of the nutrients were released from dietary plant fibres in the rumen. Absorption of water, minerals, some nitrogen, volatile fatty acid and carbohydrate occurs in this region.

The terminal end of the gastrointestinal tract is the rectum. Its function is involved with defecation or elimination of the materials remaining after passage through the digestive system.

Non-ruminant Herbivore Digestive Tract

The digestive system of the non-ruminant herbivore such as the horse, rabbit and guinea pig combines features of both the ruminant and monogastric systems. The proximal part (stomach, small intestine) of the gastrointestinal tract is very similar to that of the monogastric. The digestion of the dietary fractions which would also be available to the monogastric (i.e., the non-structural carbohydrates such as starch, proteins, minerals and vitamins) takes place in the stomach and small intestine. The more fibrous part of the feed, which in some situations is the majority of the diet, is passed through the small intestine to the hindgut (caecum and colon) where it undergoes digestive processes similar to those occurring in the rumen. The bacteria which inhabit the caecum and colon of the horse are, in most cases, identical to those found in the rumen of cattle. The end products of bacterial fermentation are also very similar to those of the ruminant animal. Since the major site of bacterial fermentation in the hind-gut fermenters is distal to the small intestine the non-ruminant herbivore may not be as efficient in using fibrous foodstuffs as is the ruminant. Much of the bacterial protein and cell wall constituents which would be available to the ruminant are lost to the horse.

There is absorption from the hind-gut of by-products of bacterial fermentation such as volatile fatty acids, free amino acids, B-vitamins and minerals released from plant sources, but losses of these nutrients in the faeces is greater in the horse than in the cow. The rate of passage of ingesta through the gastrointestinal tract of the horse is, however, more rapid than in the ruminant. This allows the horse to meet a great part of its requirements from fibrous sources by ingesting large quantities, digesting what is available and eliminating the remainder from the tract relatively quickly. For horses with high nutrient requirements (i.e., pregnant, high performance and working horses) a diet with a high proportion of cereal grains and supplements is necessary to meet their nutrient requirements.

Monogastric Digestive Systems

In animals like pigs, the lips, cheeks, palate and tongue are all involved with prehension and movement of feed in the mouth. Small feeds are usually ingested with the tongue, while the teeth shear off any large or long pieces. Pigs can also suck up slurry feeds and water. Saliva is produced by three primary glands, and its volume and consistency can change with the nature of the food. Once food has been chewed and mixed with saliva to a proper consistency it is swallowed and passed down the esophagus to the stomach. The motility of the stomach is necessary to mix the digesta with the gastric juices and to move the digesta into the small intestine.

The small intestine, comprised of the duodenum, jejunum and the ileum, is the site where the majority of digestion takes place and most, if not all, nutrient absorption occurs. The duodenum is the site for the mixing of digesta with intestinal, liver and pancreatic secretions. These secretions serve to buffer the contents as they leave the stomach, and to lubricate the bolus for ease of movement through the intestines.

The caecum and colon (hind gut) retrieve any nutrients, primarily water and electrolytes, remaining in the digesta as it leaves the small intestine. The caecum is a blind sac arising at the junction of the ileum and colon. Anaerobic fermentation of fibre in the caecum and colon produces some utilisable energy in the form of volatile fatty acids. The amount of energy produced is small in relation to the pig's total requirement but hind gut fermentation liberates substantial nutrients in horses and rabbits.

Avian Digestive Tract

The avian oesophagus has an outpocket, the crop, which occurs about two-thirds of the way down its length, right before it enters the thorax. There are large numbers of mucous glands above the crop which serve to augment the small amount of saliva produced. The crop functions mainly to store feed. From the oesophagus, the bolus moves into the proventriculus, where HCl and pepsinogen are produced at higher levels per unit body weight than most mammals. The gizzard, a very hard, muscular organ is the site for grinding of the ingesta and also for much of the gastric digestion. Movement of ingesta is related to a coordinated effort of the crop, lower oesophagus and gizzard, with little muscular contraction of the proventriculus involved.

The level of fibre in broiler rations has an effect on the size of the proventriculus and gizzard. The more fibre, the slower the rate of passage through the proventriculus. Thus the amount of gastric juices produced is less, the more the gizzard must grind the feed, so the larger the gizzard. The gizzard further degrades the ingesta to prepare for digestion in the small intestine. The ingesta is mixed in the gizzard and when the particles are fine enough they are pushed out into the small intestine. Intestinal contents are continually refluxed back into the gizzard which differs from most animals since the pylorus effectively blocks back-flow of fluid from the small intestine into the stomach. Grit was thought to be necessary for the digestion of coarse feedstuffs, but there seems to be no advantage to adding grit to a mash diet.

The small intestine, about 125 cm long in a mature chicken, is made up of the duodenum, jejunum and the ileum. The duodenum ends where the bile and pancreatic ducts empty into the intestine. The jejunum and the ileum cannot be readily differentiated. The large intestine absorbs some of the nutrients, usually water and electrolytes, remaining in the digesta as it leaves the small intestine. The colon in avian species is very short and the ceca are much longer and are paired. Only fluids and very fine particles enter the ceca and once they are evacuated from the ceca are rapidly excreted. Volatile fatty acids are by-products of anaerobic fermentation of fibre in the ceca.

Ducks and geese have longer digestive tracts than do chickens. Thus, these species can be productive on diets that are higher in fibre. The energy derived from volatile fatty acid production is small compared to the chickens' total requirement but can be substantial in ducks and geese. The colon acts more as a passage of ileal and cecal digesta than as a fermenter or absorber. The cloaca is the terminal end of the gastrointestinal urinary and reproductive systems. Voiding moves faeces and uric acid out of the cloaca together.

Other Major Organs of Digestion

The pancreas and liver are two other organs intimately associated with digestion. In addition to producing several hormones, the pancreas secretes a number of digestive enzymes such as lipase, amylase, trypsinogen, chymotrypsinogon that are essential for breaking down complex feed components and liberating the nutrients in forms that can be absorbed. The liver adds bile to the ingesta as it passes through the small intestines. The bile emulsifies fats so they become soluble and detoxifies potentially harmful materials that gain entrance to or are formed within the body.

Digestion and liberation of nutrients from animal tissues occurs quickly and efficiently during passage through the stomach and intestines. In contrast, digestion of complex plant tissue requires retention of ingested feeds for a considerable period under conditions suitable for microbial fermentation. The gastrointestinal anatomy and related ability to promote fermentation allow classification of animals into general groups as shown in the following table.

The breakdown of lignocellulose in the rumen and other parts of the digestive system is limited by the crystallinity of cellulose fibres and the barrier to microbial attack created by the lignin and other related components of the mature plant wall. Various fungi with high lignin degrading activity are currently under study to increase understanding of the physical and enzymatic mechanisms employed in these processes. On the basis of this knowledge, solid-substrate fermentation technologies have been developed and significant improvement of in vitro digestibility can now be achieved through use of fungi carefully selected for activity on particular substrate. A specific gene coding for lignin peroxidase has been isolated, cloned, incorporated into and expressed by bacteria, so genetic engineering to construct strains with increased in vitro lignin-degrading potential, and subsequent introduction of these into the rumen, may be feasible in the future.

Crude plant by-products are abundant throughout the world, with annual production of cereal grain straw, corn stover and sugar cane tops and bagasse each accounting for many millions of tonnes. However, since these all contain high proportions of lignin protected hemicellulose, and cellulose, digestibility is low even in ruminants. Considerable research is being conducted on developing techniques that improve digestibility of low quality roughage (grinding, chopping, pressure, explosion, irradiation, chemical treatments, etc.) and successful methods to increase the digestibility would contribute dramatically to better livestock nutrition.

LABELING AND REGULATION OF ANIMAL FOOD

In the United States, all pet food is regulated by the Food and Drug Administration (FDA), the United States Department of Agriculture (USDA), and the Federal Trade Commission (FTC). It is further regulated at the state level. The Association of American Feed Control Officials (AAFCO) establishes standards on which states base their feed laws and regulations, but itself has no regulatory authority.

Dog and cat foods labeled as "complete and balanced" must meet standards established by the AAFCO either by meeting a nutrient profile or by passing a feeding trial. Cat and dog food nutrient profiles were last updated by the AAFCO's Feline Nutrition Expert Subcommittee (1991-1992) and the Canine Nutrition Expert Subcommittee (1990-1991), respectively. The updated profiles replaced the previous recommendations set by the National Research Council (NRC). The NRC published new recommendations in 2006; AAFCO is currently (2010) in the process of updating the nutrient profiles.

Products found to be "complete and balanced" by feeding trials bear the label statement "animal feeding tests using AAFCO procedures substantiate that the name of product which provides complete and balanced nutrition." The protocol requires six of eight animals complete a 26-week feeding trial without showing clinical or pathological signs of nutritional deficiency or excess. The cats' or dogs' general health is evaluated by a veterinarian before and after the test. Four blood values (hemoglobin, packed cell volume, serum alkaline phosphatase and serum albumin) are measured after the trial, and the average values of the test subjects must meet minimum levels. No animal is allowed to lose more than 15% of its starting weight.

Products formulated with ingredients to meet the established nutrient profile would include the following statement, "(Name of product) is formulated to meet the nutritional levels established by the AAFCO Cat/Dog Food Nutrient Profiles." There are two separate nutrient profiles - one for "growth and reproduction" and another for "adult maintenance". The nutritional adequacy statement would include information on the life stage(s) for which the product is suitable. A product labeled as "for all life stages" must meet the more stringent nutrient profile for "growth and reproduction". Products labeled as "intended for intermittent or supplemental feeding" do not need to meet either profile.

The "Family Rule" allows a manufacturer to have a product that is "nutritionally similar" to another product in the same "family" to adopt the latter's "complete and balanced" statement without itself undergoing any feeding tests. The "similar" food must be of the same processing type; contain the same moisture content; bear a statement of nutritional adequacy for the same or less demanding life stage as the lead product; contain a dry matter, metabolisable energy (ME) content within 7.5% of the lead product's dry matter; meet the same levels of crude protein, calcium, phosphorus, zinc, lysine, thiamine and for cat foods, potassium and taurine as the lead food; and meet or exceed the nutrient levels and ratios of the lead family product or the AAFCO nutrient profiles, whichever is lower. The label statement on the similar food can be the same as the lead product if the ME is substantiated by the 10-day ME feeding study.

Critics of the AAFCO standards argue that such requirements are too lax. Generational studies conducted by researchers at University of California, shown some foods that pass AAFCO's feeding trials are still not suitable for long-term use and estimated that of 100 foods that pass the nutritional analysis, 10 to 20 would not pass the feeding trials. Although maximum levels of intake of some nutrients have been established because of concerns with overnutrition, many still lack a maximum allowed level and some contains large disparity between maximum and minimum values. The NRC accepts that despite ongoing research, large gaps still exist in the knowledge of quantitative nutritional information for specific nutrients. Some professionals acknowledge the possibilities of phytochemicals and other vital nutrients that have yet to be recognised as essential by nutritional science. With such broad guidelines and loose feeding trial standards, critics argue that the term "complete and balanced" is inaccurate and even deceptive. An AAFCO panel expert has stated that "although the AAFCO profiles are better than nothing, they provide false securities."

Certain manufacturers label their products with terms such as premium, ultra premium, and holistic. Such terms currently have no official definitions. The AAFCO is currently considering defining some of the terms. However, the terms "natural" and “organic” do have definitions; e.g., organic products must meet the same USDA regulations as for organic human food.

In Canada. products that pass the Canadian Veterinary Medical Association (CVMA) Pet Food Certification Program, which involves a feeding trial, carries a CVMA label on their packaging. Participation in the program is voluntary. There are no government regulation of pet food in the Canadian marketplace. The CVMA Program was discontinued at the end of 2007.

In the European Union, pet food is regulated by the same harmonised standards across the EU, via the Feeding Stuffs Act. All ingredients used for pet food has to be fit for human consumption according to EU requirements. But regulations require that pet food that contains by-products be labeled as “Not for human consumption” even though such by-products which have to be derived from animals declared fit for human consumption. Raw pet food has to be labeled “Pet food only”.

Products meant for daily feeding are labeled “Complete feedingstuff” or “complete petfood” or other EU languages equivalent. Products meant for intermittent feeding are labeled as “Complementary feedingstuff or “complementary pet food” while products with an ash content of over 40% are labeled “Mineral Feeding stuff”. Ingredients are listed in descending order by weight.

4

Milk Secretion

The defining characteristic of the class Mammalia is the ability to produce milk, an externally secreted fluid designed specifically to nourish the young. The provision of milk frees the mother from the necessity of providing a specialized environment for rearing of the young. It allows birth to occur at a relatively early stage of development and provides a time of intense maternal interaction with the newborn during early behavioral development. In addition the nutritional reserves of the mother may be able to sustain the suckling through a period of famine. Milk composition is species specific. In humans it is becoming increasingly clear that breast milk is the most appropriate source of nutrition up to the age of 6 months. Many components of human milk including, but not-limited to the protein lactoferrin, growth factors, long chain polyunsaturated fatty acids, bile salt stimulated lipase, and anti-infectious oligosaccharides and glycoconjugates are not duplicated in the cow's milk upon which formula manufacture is based. This article is primarily directed to a description of milk secretion in women, but material from studies in animals is drawn on as necessary to define the mechanisms involved.

MILK AS A FLUID

Milk is a complex fluid composed of several phases that can be separated by centrifugation (Neville, 1995a) into a cream layer, an aqueous phase, and a two-phase pellet. The upper phase consists of milk cells and membranous debris, the lower of casein micelles. Casein can also be precipitated by micelle-destroying treatments such as the enzyme rennin or low pH leaving an aqueous phase that is often termed whey. If the whey is made from skim milk it is a true solution that contains all the milk sugar as well as the major milk proteins lactoferrin and secretory immunoglobulin A (sIgA), the monovalent ions sodium, potassium and chloride, citrate, calcium, free phosphate and most of the water-soluble minor components of milk. Depending on the species, varying proportions of the lactoferrin, lysozyme, citrate and calcium may be found associated with the casein pellet . The casein fraction is a small proportion of human milk, about 0.2% by weight. However, casein makes up 4% of cow's milk and as much as 12% of rodent milks. The casein fraction from cows milk, usually obtained by rennin precipitation, is used in cheese-making while the whey finds a multiplicity of uses, most notably as the base for infant formula.

Functional Anatomy of The Human Breast

Nipple and coursing through the mammary fat pad to terminate in grape-like clusters of alveoli. Each duct serves a specific lobule. The lobules are separated and supported by thick connective tissue septa and, in the non-pregnant, non-lactating breast, by large amounts of adipose tissue. Blood vessels, nerves and lymphatics run in the septa which merge imperceptibly with the fascia at the anterior thoracic wall. The nipple, which serves as the termination point for the lactiferous ducts, is surrounded by an area of pigmented skin, the areola, containing sebaceous glands and sweat glands. The areola serves as the termination point for the fourth intercostal nerve which carries sensory information about suckling to the spinal cord and brain. This is extremely important in the regulation of oxytocin secretion from the posterior pituitary and prolactin from the anterior pituitary. The mammary ducts expand slightly to form sinuses beneath the areola.

During pregnancy the alveolar complexes increase in number and complexity and the cells lining the alveoli and small ducts mature, acquiring the capability to secrete milk. However, milk secretion is kept in check by high concentrations of circulating sex steroids, primarily progesterone. At parturition a series of programmed changes transforms the cells into the fully secretory state. This transformation is termed lactogenesis.

The Mammary Alveolus

Although the location and external form of the mammary glands differ from one species to another, the mechanisms of milk production are remarkably similar. Milk is produced and stored in alveolar units. Removal of milk from the alveoli is accomplished by contraction of the myoepithelial cells surrounding the alveoli (a) and ducts (d). This process is called milk ejection. Milk exits through ductules into ducts draining several clusters of alveoli. In the human the small ducts coalesce into 15 to 25 main ducts that drain sectors of the gland. The main ducts dilate into small sinuses as they near the areola where they open directly on the nipple. In many species including both ruminants and rodents the ducts empty into a single primary duct or a cistern which in turn is drained by a single teat canal. In dairy animals the cistern provides additional milk storage.

In comparison to related dermal glands such as the salivary and sweat glands, the rate of milk secretion is slow, about 1.5 ml of milk per gram of tissue per day (Peaker, 1977). Histologically, the cells lining the smaller ducts resemble the alveolar cells, even reacting with anticasein antibodies. The larger ducts play a passive role in milk secretion, merely transferring the milk from the alveolar stores to the sub-areolar sinuses where it becomes available to the suckling infant. Because the composition of the aqueous phase of milk changes very little during a feed or milking, it is unlikely that re-absorptive processes, like those important in the formation of saliva or sweat, play a significant role in determining milk composition.

Although the mammary epithelial cells are ultimately responsible for converting most precursors into milk constituents and transporting them to the mammary lumen, other cell types are also intimately involved in milk production. We have already mentioned the myoepithelial cells responsible for milk ejection from the breast.

The mammary ducts and alveoli are embedded in a stroma that contains fibroblasts, adipocytes, plasma cells and blood vessels. Blood flow is greatly expanded during lactation to make available the large amounts of substrate required for milk synthesis. Interactions with stromal cells are intimately involved in mammary development and milk secretion. Stromal fibroblasts and\or adipocytes are known to be the source of growth factors such as hepatic growth factor/scatter factor and IGF-1 and are probably responsible for production of the enzyme lipoprotein lipase, important in milk lipid synthesis. During lactation B lymphocytes "home" to the mammary gland where they become plasma cells and settle in the interstitial space producing the immunoglobulins that ultimately find their way into milk.

Five distinct processes are utilized by the mammary epithelium in the secretion of milk. These pathways operate in parallel to transform precursors derived from the blood or interstitial fluid into milk constituents. Although the biochemical processes involved are fundamentally the same in all mammals, differences in their relative rates and, in some cases, in the nature of the products synthesized result in milks whose composition differs widely from species to species. Some of the milk secretion pathways, e.g., exocytosis of protein-containing vesicles and transcytosis of immunoglobulins, are similar to processes in many exocrine organs. In contrast, the mechanism for fat secretion is unique to the mammary gland.

CELLULAR MECHANISMS FOR MILK SYNTHESIS AND SECRETION

Four secretory processes are synchronized in the mammary epithelial cell of the lactating mammary gland: exocytosis, lipid synthesis and secretion, transmembrane secretion of ions and water and transcytosis of extra-alveolar proteins such as immunoglobulins, hormones and albumin from the interstitial space. A fifth pathway, the paracellular pathway, allows the direct transfer of materials between the milk space and the interstitial space. This pathway is open in the pregnant gland and allows the transfer of molecules at least as large as intact immunoglobulins. It is closed in the fully lactating gland providing a tight barrier between the milk and interstitial spaces. This barrier opens again in the presence of mastitis and during involution.

Exocytosis

Most of the components of the aqueous phase of milk are secreted by the exocytotic pathway. Proteins synthesized on ribosomes are transferred to the lumen of the rough endoplasmic reticulum where their signal sequences are cleaved and the protein molecules folded. Vesicles transfer the proteins to the Golgi stack where they are further processed by the addition of carbohydrate, phosphate or other groups and packaged into secretory vesicles.

In addition to processing of milk proteins per se, the Golgi vesicles in the lactating mammary cell synthesize lactose from precursor UDP-galactose and glucose that enter from the cytoplasm. Because the Golgi membrane is impermeable to lactose, the sugar is osmotically active and water is drawn into the terminal Golgi vesicles. The swollen appearance of the trans-Golgi and the secretory vesicles which arise from it are specific characteristics of the lactating mammary cell. Casein micelle formation begins in the cis-Golgi with condensation of casein molecules; addition of calcium, possibly in the secretory vesicle, leads to maturation of the casein micelles into particles sufficiently dense to be seen in the electron microscope. Secretory vesicles are thought to be the source of most of the constituents of the aqueous phase of milk including citrate, nucleotides, calcium, phosphate and probably monovalent ions and glucose. However, the apical membrane of the mammary alveolar cell has transporters for monovalent ions and glucose and the concentration of these substances may be adjusted by direct membrane transport.

Secretory vesicles move to the plasma membrane where they fuse and release their contents into the milk space by exocytosis. The exocytotic pathway involved in milk secretion appears to be largely constitutive. This means that, once secretion begins after parturition, exocytosis is continuous and secretory products are not stored within the epithelial cell. However, Burgoyne and his colleagues have observed that a portion of the mammary secretion could be stimulated by increased cell calcium in an in vitro system suggesting the presence of a regulated secretion pathway in the lactating mammary epithelial cell.

Lipid Synthesis and Secretion

Triglycerides, synthesized in the smooth endoplasmic reticulum of the mammary alveolar cell from precursor fatty acids and glycerol, coalesce into large droplets that are drawn to the apex of the cell. The lipid droplets bulge against and gradually become enveloped in apical plasma membrane, finally separating from the cell as the milk fat globule. The occasional inclusion of a crescent of cytoplasm within the membrane-bound globule enables any substance contained in the cytoplasm to enter milk. The membrane surrounding the milk fat globule has two functions: it is the primary dietary source of phospholipids and cholesterol for the breast-fed infant and it prevents the fat globules from coalescing into large fat droplets that might prove difficult to secrete..

Transport across the Apical Membrane

In contrast to the other pathways for milk secretion the pathways for the direct transport of substances across the apical membrane of the mammary alveolar cell are poorly understood. Linzell and Peaker devised a clever technique to determine what molecules could utilize this pathway, infusing isotopes of small molecules up the teat of a goat and calculating how much of the substance left the milk and entered the blood. They found that sodium, potassium, chloride and certain monosaccharides as well as water directly permeated this membrane but calcium, phosphate and citrate did not. Studies of bicarbonate secretion led these investigators to postulate the presence of chloride-bicarbonate exchange at the apical membrane. Stable isotope studies in women confirmed the presence of a glucose pathway across the apical membrane in the human mammary gland as well. The function of the apical pathways is not clear, since the composition of the aqueous phase of milk is thought to be determined in the Golgi and secretory vesicles. What is clear, however, is that apical pathways are limited to a modest number of small molecules. Although often overlooked, many drugs enter milk by direct transfer across both basolateral and apical membranes of the mammary alveolar cell. For this reason most therapeutic drugs are transferred efficiently into milk.

Transcytosis of Interstitial Molecules

Intact proteins can cross the mammary epithelium from the interstitial fluid either by transcytosis or through the paracellular pathway. During lactation only the transcytotic pathway is available. Immunoglobulins are the best studied of the molecules that enter milk via transcytosis. In most-non-ruminants IgA is synthesized by plasma cells in the interstitial spaces of the mammary gland or elsewhere in the body. The protein binds to receptors, the polymeric immunoglobulin receptor, on the basal surface of the mammary alveolar cell; the entire IgA-receptor complex is endocytosed and transferred across the cell. At the apical membrane the extracellular portion of the receptor is cleaved and secreted together with the IgA. The cleaved receptor portion is known as secretory component and the secreted product is thus secretory IgA or sIgA.

The Paracellular Pathway

The paracellular pathway allows passage of substances between epithelial cells, rather than through them. During full lactation the passage of even small molecular weight substances between alveolar cells is impeded by a gasket-like structure called the tight junctions that joins the epithelial cells tightly, one to another. Although immune cells apparently can diapedese between epithelial cells to reach the milk, the junctions seal tightly behind them leaving no permanent gap. During pregnancy, with mastitis and after involution the tight junctions become leaky and allow components of the interstitial space to pass unimpeded into the milk. At the same time milk components can enter the plasma. This leakiness is useful during these periods since secretion products are allowed to leave the gland, inflammatory cells and protective molecules can enter the milk space and products of the dissolution of the mammary cells during involution can be cleared from the breast.

LACTOGENESIS: THE TRANSITION FROM PREGNANCY TO LACTATION

Formally lactogenesis is defined as the onset of milk secretion. Hartmann (1973) and Linzell and colleagues, based on work in ruminants, divided lactogenesis into two stages. Lactogenesis Stage 1 occurs during pregnancy when the gland becomes sufficiently differentiated to secrete small quantities of specific milk components such as casein and lactose. Lactogenesis Stage 2 is defined as the onset of copious milk secretion associated with parturition. It is brought about by a decline in progesterone around the time of parturition in the presence of maintained prolactin concentrations. A differentiated mammary epithelium is necessary for stage 2 lactogenesis to occur. In humans the epithelium reaches this stage of differentiation about mid-pregnancy.

In the early post-partum period the secretion product of the mammary gland is called colostrum. This fluid contains high concentrations of immunoglobulins and the protective protein, lactoferrin. In species such as ruminants that lack transplacental transport of immunoglobulins, feeding of colostrum is necessary to provide passive immune protection to the young until their immune systems become mature. In other species, such as humans, where transplacental transport of immunoglobulins provides humoral immunity in the early post-partum period, the presence of secretory IgA, lactoferrin and high concentrations of oligosaccharides is important in protection of mucosal surfaces from infection, particularly under conditions where optimal sanitation cannot be maintained.

Lactogenesis represents a profound and rapid series of changes in the activity of differentiated mammary epithelial cells from a quiescent state to a fully active secretory state. As can be illustrated by examining the composition of human milk during the first week post-partum, these changes occur as an orderly progression of events that starts with closure of the tight junctions between the epithelial cells, followed by a transient increase in the secretion of the protective proteins sIgA and lactoferrin. After about 36 hours a rapid increase in the synthesis of all the components of mature milk begins that is complete by about day 5 postpartum

The first change to occur is a fall in the sodium and chloride concentrations in the milk and an increase in the lactose concentration. These modifications commence immediately after birth and are largely complete by 72 hours postpartum (Neville et al., 1991). They precede the onset of the increase in milk volume by at least 36 hours and can be explained by closure of the tight junctions. With blockage of the paracellular pathway lactose, made by the epithelial cells, can no longer pass into the plasma. Sodium and chloride can no longer pass from the interstitial space into the lumina of the mammary alveoli. Thus this phase is marked by a rapid fall in the sodium and chloride concentrations of milk and a rapid increase in the lactose concentration.

The next change is an increase in the rates of secretion of sIgA and lactoferrin. The concentrations of these two important protective proteins remain high for the first 48 hours after birth, together comprising as much as 10% by weight of the milk. Colostrum also contains high concentrations of cells including lymphocytes, macrophages, neutrophils and sloughed secretory epithelium. The concentrations of lactoferrin, sIgA and cellular elements fall rapidly after day 2, a consequence of both dilution as milk volume secretion increases and a decrease in their absolute rate of secretion. Although both lactoferrin and IgA are found at high concentrations in colostrum, their secretion is likely under separate control since lactoferrin is secreted by the exocytotic pathway while sIgA is secreted by the transcytotic pathway.

Finally, starting about 36 hours postpartum there is a 10-fold increase in milk volume from about 50 ml/day to 500 ml/day. This volume increase is perceived by the parturient woman as the "coming in" of the milk and is brought about by a massive increase in the rates of synthesis and/or secretion of almost all the components of mature milk (Neville et al.1991), most notably lactose, casein (Patton et al., 1986), alpha-lactalbumin, lipid, calcium, potassium, etc. Three milk components, citrate, glucose and free phosphate, are quite interesting because their concentrations increase in proportion to the increase in milk volume In the case of glucose the change in concentration has been shown to be due to an increase in glucose transport from the interstitial space into the mammary alveolar cell (Neville et al.,1990) whereas the increase in phosphate possibly results from dephosphorylation of the UDP generated during lactose synthesis. The increase in citrate may be related to fat synthesis; however, it does not occur in all species since rodents do not have citrate in their milks.

Acute heat stress brings down milk secretion in dairy cows by up-regulating the activity of the milk-borne negative feedback regulatory system

In hot climates, high ambient temperatures, and high direct and indirect solar radiation, wind speed and humidity, are the main environmental stressing factors that impose stress on animals. Cattle have a higher metabolic rate than most other domestic ruminants, and a poorly developed water retention mechanism in the kidney and gut. Furthermore, as a consequence of aggressive selection for milk production over the last 5 decades, modern cows in Israel and the USA produce 40–70 L of milk per day, compared with 10 L day-1 or less in their ancestors. Each 10 L day-1 of milk yield roughly doubles the metabolizable energy requirement of cows, and ~35% of this energy is dissipated as heat. High-yielding cows are affected more than low-yielding ones because the upper critical temperature shifts downward as milk production, feed intake, and heat production increase.

Acclimation involves phenotypic responses to environmental changes, which are reflected in hormonal signals and also in alterations in target tissue responsiveness to hormonal stimuli. The time required for acclimation varies according to tissue types, and ranges from a few days to several weeks; for example, changes in metabolism in response to HS occur over a few days .

However, this picture does not cover all situations; some external environmental stresses, such as dehydration and acute HS may very quickly (within 24 to 48 h) take animals beyond their current acclimatized-adaptive range, which necessitates the induction of emergency physiological responses in order to avoid sudden death. Such immediate measures include an acute reduction in milk yield, because milk production, particularly in high-yielding dairy cows, intensifies the effects of these external stresses.

Milk secretion and mammary function are regulated acutely by local autocrine feedback mechanisms that involve milk-borne factors which are sensitive to the frequency and efficiency of milking ,Sustained changes in the frequency of milking and milk secretion are associated metabolic adaptations and with longer-term adaptations in the degree of differentiation and, ultimately, the number of mammary epithelial cells. Consequently, such changes may affect the above-described homeostatic processes. It has been hypothesized that the fast modulation of milk secretion in response to external factors, such as emotional stress, and harsh physical conditions such as heat stress and water deprivation also depends on such a negative-feedback regulatory system, which increases the potential for survival in response to stress.This negative feedback system was shown to comprise an endogenous milk enzymatic system, the plasminogen activator (PA)-plasminogen (PG)-plasmin (PL), that specifically forms a â-casein (CN) fragment (f) (1–28) from â-CN, which acts as the negative control signal by closing potassium channels on the apical membrane of the epithelial cells of the mammary gland .Down-regulation of these channels induces undefined inwardly directed cellular signals that inhibit milk secretion. Interestingly, a further activation of the PA-PG-PL system, which was coupled with more extensive degradation of casein induced involution of the mammary gland in lactating goats and cows and forcefully activated the innate immune system. Based on these findings, a casein hydrolyzate preparation was developed to reduce the suffering from mammary gland engorgement associated with abrupt cessation of milking (the conventional procedure to induce involution in modern dairy cows) and to treat and prevent common clinical and subclinical infections of the udder in dairy cows.

The concept that PA-PG-PL-â-CN f(1–28) is involved in milk-born negative feedback regulation of milk secretion was supported experimentally under conditions that simulated stress, and by exposure of cows to dehydration

Methods

Ethics

All protocols were approved by the Institutional Animal Care Committee of the Agricultural Research Organization, which is the legitimate body for such authorizations in Israel.

Study layout

The experiment was carried out at the height of summer (late July) in the experimental dairy herd of the Agricultural Research Organization, at the Volcani Center at Bet Dagan. Eighteen Israeli Holstein cows, between their second and fourth lactations, with milk yields of 48–52 l day-1 were subjected to the study. The cows were allocated to three treatments, each of six cows, according to lactation number and milk yield (MY). The treatments were: treatment C – the cows had access to shade and benefited from the conventional cooling procedures in the herd; treatment D – the cows were denied access to shade, but benefited from the cooling procedures; and treatment E – the cows were denied access to both shade and cooling procedures. The cows were held in small yards; those in treatment C had 10 m2 of a shaded slatted floor and 10 m2 of an unshaded concrete – surfaced yard per animal, whereas those in treatments D and E had 20 m2 of an unshaded concrete – surfaced yard per animal. The cows were fed before and during the experiment with a typical Israeli total mixed ration (17% protein) containing 65% concentrate and 35% forage, which was offered ad lib in managers; water was available at all times. At the noon milking before the start of the experiment, milk samples were taken from all cows, and then the cows were allocated to treatments.

The cows were milk three times daily (0530, 1230, and 2130), and the and exact milking times were individually recorded automatically.The cooling system has been described previously. Briefly, an array of fans produced air velocities of 2 m/second or more from 0600 to 2400. In addition, fans and sprinklers were sequentially activated to repeat cycles of wetting (0.5 min) and ventilation (4.5 min) for seven 0.5-h periods every 1.5 to 2 h, between 0730 and 1830.

Ambient conditions

Air temperature and relative humidity were recorded at a meteorological station located 1.5 km from the farm. Average maximal (noon) temperature during the three days of the experiments were 31, 35 and 37°C, minimal (night) temperatures were 18, 20 and 22°C, relative humidity was 81, 85 and 88%; the temperature humidity index values at mid-day were 80, 82 and 88.

Rectal temperature and respiration rate

Respiration rates were determined at 1300 by visually measuring the breathing rates with the aid of a stopwatch. Rectal temperature was measured at 1400 to nearest 0.1°C, with an interchangeable thermistor probe system (model 46TUC Tele-thermometer, Yellow Springs Instruments, Yellow Spring, OH)

Gross composition, plasmin activity, Na+ and K+ concentrations

Gross composition, the activities of PA, PG, and PL and the concentrations of Na+ and K+ in the milk were determined essentially as described previously.

K+ uptake into skim milk vesicles

Milk was fractionated into clear infranatant (Inf; i.e., milk serum devoid of vesicles and casein), skim-milk-derived vesicles (SMV) and casein by a combination of centrifugation and ultracentrifugation .The Inf and SMV were stored at -80°C prior to analysis. The clear Inf was used for measuring K+ channel blocker activity by comparing Rb uptake into SMV with those obtained after incubation of the

SMV in sucrose solution. Protein concentrations in Inf and SMV preparations were assayed by the Bradford method.

The procedure for assaying K+ uptake into the SMV was originally described by Shennan and applied for assessing the activity of the milk-borne potassium channel inhibitor, as described previously,; briefly, uptake at 23°C, was initiated by adding 10 ìl of RbCl to a reaction medium consisting of 25 ìl of SMV, 325 ìl of the working solution (control) or 300 ìl of working solution, and 25 ìl of test samples. Disposable Pasteur pippets, plugged with polymer filter wool and filled with a cation exchange resin (Dowex, hydrogen form, 50–100 dry mesh, 8% cross-linked), were used to separate intravesicular from extravesicular isotope. Four minutes after the addition of RbCl, a 200-ìl sample of the reaction mixture was applied to the Dowex column, which was then washed with 2 ml of an ice-cold solution that was collected in scintillation vials. The radioactivity of the samples was counted in a Liquid Scintillation Counter (Tri Carb, Hewlett-Packard).

Preparation of â-CN f (1–28) and measurement of its K+ channel blocking activity â-CNf (1–28) was selectively precipitated from a plasmin digest of â-CN by using acidified calcium chloride and ethanol.The peptide was further purified by anion exchange fast protein liquid chromatography and the preparative mode of reversed-phase HPLC, which was also used to collect the peptide .The identity of the major peak collected, which amounted to ~95% of the total peptides, was verified by matrix-assisted laser desorption ionization time-of flight mass spectrometry at the Technion, Haifa, Israel. The K+ uptake into SMV was by comparing SMV derived from the morning milking of control cows and the noon milking of experimental cows, that contained â-CN f (1–28) added at 5 ìg ml-1 (~1.5 ìM), with a mixture that contained only the incubation solution.

Statistical analysis

The data were statistically analyzed with the Fit Model procedure for repeated measurements, within the JMP software, Version 5 (SAS Institute, Cary, NC, USA); the between-subject factor was the treatment, and the within-subject factor was time. Differences were considered significant at P < 0.05. The model was:

$$Y_{ijklm} = \mu + \rho_i + \alpha_j + C_{(ij)k} + \gamma_l + \alpha\gamma_{il} + \varepsilon_{ijklm}$$

where Yijklm = the dependent variable, ì = overall mean, ñi = fixed effect of period (pre- or post-treatment; i = 1 or 2), áj = fixed treatment effect j (j = 1 or 2), C(ij)k = random effect of cow k (k = 1 to 18) within period i and treatment j; al = effect of day l (l = 1 to 3 or 1 to 9 for milk yield); agil = effect of interaction of treatment j and day l; and eijklm = random error associated with cow k in period i and treatment j on day l

Differences between treatments for period, or for specific days following treatment were subjected to Student's t-test by means of the Tukey-Kramer Highest Significant Difference (HSD) test.

Results

Respiration rate per minute measured on the second day of the experiment at 1300 varied significantly among the treatments being 68 ± 7 in treatment C, 102 ± 8 in treatment D and 152 ± 10 in treatment E ($P < 0.01$). Similar ranking (39. 1 ± 1, 39.5 ± 1, 41.6 ± 1, respectively; $P < 0.01$) were recorded with respect to rectal temperatures measured at 1400 on the second day.

Milk yields of all cows were in the range 48–49 l/d before the start of the experiment, and did not differ between the treatments at that time. All treatments induced a significant drop in milk yield, starting at day 1, which maximized at day 2 of the experiment. However, the treatments differed in the extent of the responses: treatment E (a maximal drop of 55.1% in comparison to the pre-treatment yield) > treatment D (-18.2%) > treatment C (-7.9%). After returning the cows to their original group in the afternoon of day 2, treatment's effects on milk yield faded within the next 4 days, the recovery appeared to be slower (within 7 to 10 days) in treatment E.

Hourly milk secretion rates were determined by dividing the milk recorded at milking by the time interval (~8 h) since the previous milking. Though, within each group, the yield tended to be higher between 0600 and 1400 than between 1400 and 2200, the figures did not differ statistically. Therefore, we pooled the milk yields into night-time yield (2130 to 0530 the next morning) and daytime yield (0530 to 2130 on the same day). Within each group, the milk secretion rate was higher during the night than during the day according to treatments, E difference > treatment D difference = treatment C difference; the daytime yields being lower than the night-time ones by 48, 20 and 19%, respectively. Consequently, the difference in milk yield between treatment E and treatments D and C was smaller during the night than during the day.

The treatments did not induce changes in concentrations of lactose, Na (~16 mM), and K (~43 mM), nor,, in the Na/K ratio (~2.7), but elicited increases in total fat and protein concentrations. The increases in fat and protein concentration were more apparent in the samples taken at 1400 than in those taken at 0600 am, and the treatments differed in the magnitude of this response, in the order, treatment E > treatment D > treatment C. The increases in fat and protein concentrations elicited by the treatments did not compensate for the drops in their yields caused by the fall in milk yield; however, the differences between night and day secretions of fat and protein were not observed in treatment D.

Effect of treatments and time of sampling on K+ uptake into vesicles derived from milk serum and incubated with the respective infranatant (milk serum devoid of vesicles and casein micelles). All the treatments (C, Control, D, treatment D, E, treatment E) and sampling time within treatment (night vs. day) effects were significant at $P < 0.05$).

It depicts the results of K+ uptake determinations; it presents comparisons of all possible combinations of Inf and SMV sampled from the control cows at the morning milking with those sampled at midday from the experimental cows. The combination of Inf and SMV sampled from the morning milking of the control cows (Cinf-CSMV) served as the reference value. Inhibition of K+ uptake into the vesicles diminished in the order: Einf-ESMV > Einf-CSMV > Cinf - ESMV, with values of 52, 68 and 77% of Cinf-CSMV, respectively. The data here are consistent with those in, which shows that acute heat stress increased the concentration of K+ channel blocker in the milk serum in proportion to the heat stress level. However, the data also indicate that vesicles (SMV) coming from heat-stressed cows were more responsive to the activity of the K+ channel blocker.

Effects on K+ uptake into the vesicles, of all combinations of incubating infranatant (milk serum devoid of vesicles and casein micelles) with milk serum-derived vesicles sampled from the control cows at the morning milking and experimental cows at the noon milking. All combinations differed significantly at $P < 0.01$.

Previously, we identified â-CN f (1–28), as the blocker of the putative K+ channel in milk serum-derived vesicles, which serve as model of the K+ channel located on the apical aspect of mammary gland epithelial cells. In order to refine the finding that vesicles derived from heat-stressed cow were more susceptible to the K+ channel blocker than those from unstressed cows, we incubated vesicles derived from the morning milking of control cows and the noon milking of experimental cows, with a fixed amount (~1.5 ìM) of âCN f (1–28), and measured their K+ uptakes. The data here clearly show that vesicles derived from the experimental cows were more susceptible to the channel blocking activity of â-CN f (1–28) than those derived from the control cows. In general, the level of inhibition obtained in this experiment was consisted with previous dose-response test results.

Discussion

Our search for the involvement of apically located K+- channel in the mechanism of feedback regulation of milk secretion was inspired by the need for extensive apical K+ conductance, since K+- concentration of milk is considerably higher than that of plasma .Indeed, K+ channels have been localized in the apical aspect of mammary secretory epithelia and the present study] and have been shown to be associated in milk-born negative feedback regulation of milk secretion. The voltage-depended nature of the apical K+- channel, suggests that it belongs to the family of the voltage-gated potassium channels. Indeed, a KCNQ1 (a member of the voltage-gated K+- channels) and a number of its accessory subunits were recently localized in the human mammary epithelial cell line, MCF-7 (33). The assembly of KCNQ1 with the accessory units may shift the K+ channel to an open state, associated with hyperpolarization of the membrane potential, which in general facilitates secretary phenomena. Thus, the milk-born regulatory element (i.e., â-CN f (1–28)), which blocks apically located K+- channel ,present study] should cause depolarization of the membrane potential, and in-turn to down regulate cells metabolism and secretory activity. As discussed below, the present results further support the concept that apically located K+- channel and its natural milk-born regulator are involved in negative feedback mechanism that controls milk secretion.

The involvement of PA-PG-PL system in regulation of milk secretion and the induction of mammary gland involution are well established. The effect of the PA-PG-PL system was related to its contribution to enhanced degradation of the extracellular matrix. However, this alternative explanation fell to explain its effect on milk secretion and its sharp increase immediately after the start of active involution (i.e., involution stage I) because in both cases there is no increase in extracellular matrix degradation. The present results as discussed below in line with previous results, strongly suggest that the PA-PG-PL system works in mammary secretion by increasing casein degradation and liberation of active components, and that â-CN f (1–28) is a principal casein degradation product that involves in negative control of milk secretion.

Animals can adapt to hot environmental conditions by gradual acclimation. As the present experiment was carried out in at mid-summer, it appears logical to assume that the cows were already acclimated to heat stress. Therefore, we first examined whether the treatments induced acute HS responses in the participating cows, and whether the effects of between-treatments differences in the imposed heat loads were significant?

The temperature-humidity index (THI) is a parameter that is widely used to describe the thermal stress imposed by climatic conditions on humans and farm animals. The THI is calculated from a combination of wet- and dry-bulb air temperatures for a particular day: THI values of 70 or less are considered comfortable, 75–78 stressful, and values greater than 78 cause extreme distress. According to these criteria the cows in the three treatments were under extreme heat stress during all the days of the experiment, though the HS condition worsened quite dramatically on the second day. Nevertheless, the THI in this case obviously could not account for the differences between treatments in milk yield, in light of previous experimental findings that the THI accounted for only small proportion of the variance of heat stress-related milk yield in dairy cows . This is in part because of wide variations between individuals, in part because the animal is related to its environment in a much more complex manner than is represented by this index, and in part because of the cooling measures applied, that help to relieve some of the heat stress. A major factor that is not taken into account in the THI is solar radiation. Multiple regression analysis that used respiration rate and body temperature as the dependent response variables to derive an index of effective temperature load has shown that the constant associated with radiation (black globe temperature) was three times as great as that associated with ambient temperature.

In order to solve the above problems, thermal physiologists adopted the stress-strain relationship from physical sciences as an applicable concept, with strain referring to an internal displacement from the resting or basal state brought about by external stress. Based on the differences in respiration rate and rectal temperature, the heat-load-related strain response diminished in the order, treatment E > treatment D > treatment C. A more detailed description than that provided by the classical thermoneutral zone concept of the relation between an animal and its environment was proposed, according to which, treatment E brought the cows deep into the noxious stage, in which coping attempts to maintain the correct body temperature are markedly unsuccessful, and drove them too close to the extreme stage, which is defined as that in which a vicious cycle of rising body temperature starts that eventually may become lethal. Therefore, the experiment was terminated at this stage, and the cows were returned to their normal, most optimal conditions (with access to shade and cooling).

Lactation is a physiological process that presents a substantial challenge to the homeostasis of the cardiovascular and fluid secretory system. The acute and large decrease in milk secretion, particularly in treatment E, may, therefore, be considered as having vital importance that makes it necessary to enable the cows to survive the HS. Thus, our present results conflict with the concept that the initial reactions of the animal to acute heat stress represent merely an emotional rather than a thermoregulatory response.

During the summer months in Israel and hotter parts of the USA, milk yields, as well as milk protein fat contents, was found to be reduced, even in dairy herds in which the cows were cooled, which indicates that these reductions form part of their adaptive response. On the other hand, acute stress in response to intramammary treatment with dexamethasone or to dehydration resulted in more intense inhibition of lactose and fluid secretion than of fat and protein secretion, which was reflected in increased fat and protein concentrations in milk, though these increases did not compensate for the overall reduction in their yields. Thus, the present results regarding the effect of HS on fat and protein concentrations are consistent with the above-noted response to acute stress. In keeping with previous findings ,in this paper we present evidence that the acute reduction in milk secretion under heat stress is associated with activation of the PA-PG-PL-release of â-CN (1–28) system.

Isolation and regrouping has been shown to impose an emotional stress that is associated with reduced milk yield and activation of the systemic stress response. Thus, the significant reduction in milk yield in treatment C, may also relate to activation of a local negative-feedback control, in response to emotional stress. However, we could not test this hypothesis because of the confounding effect of the increase in HS between the start and day 2 of the experiment, which independently could have activated the negative-feedback regulatory system.

The present study highlighted two previously unreported features of this local negative-feedback regulatory system.

First, that is a very fast-responding and flexible system, which can respond to diurnal (night vs. day) changes in the ambient conditions. Such a precise level of response to fluctuating conditions may be made possible if the mammary gland cells themselves serve as a regulatory afferent component in the regulatory system. In keeping with such a hypothesis, there is evidence that local heat shock induces endogenous hyperfibrinolysis, which is equivalent to enhanced caseinolysis in the present case, by upregulation of plasminogen activators.

The hypothesis that HS induces a short-term rapid regulatory response is consistent with evidence that in lactating cows under commercial production conditions, the effects of heat stress that may be experienced under exposure to high ambient temperatures during the day appears to be ameliorated when temperatures fall at night, and that lack of a cool night-time ambient temperature intensifies the reduction in milk yield.

Secondly, we demonstrated that expression of potassium channels on the apical membrane serves as a regulatory component in the negative-feedback system, by increasing the apical-membrane K channel sensitivity to â-CN f (1–28) levels, and this, most likely, further contributes to the accuracy and efficacy of the system. A turnover with biological half-life of 7.3 h was demonstrated for a KATP channel :thus, the rapid night to day shift in the sensitivity to K+ uptake through apical-membrane-derived vesicles is theoretically possible.

THE COW'S UDDER AND MILK SECRETION

The cow's udder consists of four entirely separate quarters. The right and left halves are separated by a distinct membranous wall called the medial (central) suspensory ligament. This ligament is composed of strong elastic tissue that extends between the halves of the udder. Numerous branches (lamellae) extend from this ligament into the quarters. As the udder fills with milk it expands or stretches like a balloon, thereby making room for the milk being stored in the udder between milkings.

The elasticity of the central ligament results in the ungainly and uncomfortable appearance of the udder in high-yielding cows. Not only does the udder force the cow's hind legs apart, making walking difficult, but the middle of the udder drops further from the body than do the outer portions, thus causing milk-engorged teats to project outwards. If the central udder support is weak, then the udder will become pendulous and the teats will face outwards, causing not only difficult milking, but also increasing their susceptibility to injury and contamination by dirt and bacteria.

The front and rear quarters of the udder are divided by a very thin wall of connective tissue which is irregular in outline. This division is difficult to see from the outside.

The outer wall of the udder contains ligaments which blend into muscles in the hind quarters of the cow. The milk-secreting parts of the udder are supported by web-like connective tissue which cris-crosses the udder . Some cows may have large udders, not because they are high producers, but because of a high content of connective tissue. This allows less space for milk-secreting tissue. Udders which shrink noticeably and become flabby after milking contain little connective or scar tissue.

Each of the four quarters contains a seperate mammary gland which, in turn is composed of:-

1. Secretory tissue (alveoli)
2. A duct system (interconnecting tubes)
3. Two cisterns (storage areas)
4. A teat

The Teats

Milk is removed from each gland by the streak or teat canal which is 8 to 12 mm long. It is kept closed between milkings by a circular sphincter muscle near the tip. This muscle is important not only in keeping the milk in the gland, but also in preventing entry by bacteria.

The character of the sphincter is important to the cow's productivity. If the canal is small, or if the sphincter is unusually strong, then the cow is hard and slow to milk. At the opposite extreme (large canal, or weak sphincter), milk will leak from the teat between milkings and the udder is then open to invasion by mastitis-causing organisms.

Components of the udder (Quinn, 1980)

Teat cistern

This cavity of the teat, located just above the streak canal, stores the milk which drains from the gland. It normally holds 15 to 40 ml of milk, depending on the size of the teat.

Blood supply

The teat contains many blood vessels with small valves that maintain blood flow when the teat is massaged (by the teat cup liner). Lack of massage during milking causes blood and other body fluids to dam-up in the teat. This is painful to the cow, and she will then be inclined to kick.

Gland Cistern

The gland, or udder, cistern is located just above the teat cistern and is partially separated from it by annular folds of tissue. Although the udder cistern varies in shape and size between quarters and cows, it stores only about 500 ml of milk.

The annular folds can become the cause of milking problems if the teat cup of the milking machine is allowed to creep up toward the udder. This pinches the annular fold and stops the flow of milk. Creeping also causes irritation to the sensitive tissue of the udder and can be the cause of injury or mastitis.

Ducts

A number of large ducts (tubes) branch off from the gland cistern. These ducts branch and re-branch into smaller and smaller ducts (similar to roots of a tree) and finally into the small ductules that drain each alveolus.

The function of the duct system is to collect the milk from the secretory tissue, to store part of the milk between milkings and to transport the milk to the gland cistern.

Where the milk ducts branch-off from a larger duct, a narrowing of the tube occurs. As the lobules (milk secreting glands) fill up with milk, their weight causes them to sag downwards. This causes the narrow branch of each duct to be pinched and close. Consequently, milk does not drain from the udder until the cow is stimulated to let-down her milk.

Secretory Tissue

The main milk-producing unit is the alveolus. This is a microscopic structure, almost spherical in shape, the outer surface of which is lined with a single layer of epithelial (milk-secretory) cells and has a tiny duct to allow the milk to drain out. The alveolus thus resembles a grape with its stalk. The outer layer also contains a complex of tiny, muscle-like cells called myoepithelial cells.

The alveoli are connected by small ducts, and groups of alveoli form lobules. A group of lobules, in turn, forms the main lobe of each quarter.The udder of the dairy cow contains about 5 x 1012 secretory cells in the epithelium of the alveolar tissue.

These have a short life and are rapidly replaced. In early lactation there is a net gain in cell numbers, but after peak yield there is a gradual decline in total numbers. Each alveolus is surrounded by a network of minute blood vessels (capillaries which carry the substances from which milk is synthesized.

When the hollow part of the alveolus is filled with milk, these small muscles expand to their fullest length. During milking the muscles contract and squeeze the milk out of the alveolus.

MILK SECRETION

Source of milk constituents

Although the site of milk synthesis is the epithelial cells lining the alveoli, some milk components (vitamins, minerals, certain proteins) are not synthesized here. They are simply filtered out from the blood, through the cells into the milk. Lactose (milk sugar), fat (butterfat), and most of the proteins (mainly casein) are formed in the epithelial cells from substrates carried in the blood. More than 300 kg of blood must pass through the udder to produce 1 R of milk.

The substrates are filtered from the blood and pass into each epithelial cell (lining the alveolus) through the plasma membrane. When this membrane functions normally, it is extremely selective as to which substances can pass through. However, when the udder is infected (for example, as in mastitis), the selectivity breaks down, and higher levels of blood protein and even white blood cells gain entry to the alveolus.

Release into alveolar lumen

The processes whereby milk constituents (fat, protein etc) are released from the epithelial cells into the cavity (lumen) of the alveolus are most fascinating. Within the epithelial cell, milk fat is produced as small, well-defined droplets. The droplets enlarge by merging together as they move toward the central lumen of the alveolus. As the fat droplet emerges from the cell it becomes totally surrounded by part of the cell wall, and an actual break or rupture in the lining forms as the fat droplet is pinched-off.

Casein and lactose are also gathered together in small packages (secretory vesicles) surrounded by a membrane. As in the case of fats, these vesicles move towards the lumen of the alveolus. When the vesicle reaches the membrane lining of the epithelial cell, its outer lining fuses with that of the cell. The contents are then discharged into the alveolar cavity.

This fusion of vesicle membrane and cell membrane repairs the loss caused by escape of fat droplets. Each epithelial cell may discharge fat or protein and casein 15 to 20 times between milkings.

The molecules of proteins, milk sugar, vitamins, and minerals are uniformally distributed within the water of the milk. In contrast, the milk fat globules float in the water after entering the lumen of the alveolus. These fat droplets then collect as tiny clusters which cannot easily pass through the tiny ducts draining the alveoli. However, the water and its contents move freely along the tubes and gradually fill the cisterns of the quarters. Some of the fat is even retained within the secretory cell and will escape only when the pressure within the udder drops during milking. Consequently the first batch of milk removed from the udder contains only 1 to 2% fat while that coming out of the alveoli may reach 24% fat.

Secretion rate

The rate at which epithelial cells manufacture milk is high for the first 10 hours after milking. It then starts to decrease at a rapid rate, and if the cow is not milked it will cease about 35 hours after the last milking. The milk secretory process stops when the pressure inside the alveolus becomes prohibitive. This probably occurs because the pressure constricts the blood vessels, and stops the flow to the milk secreting cells.

Lower-producing cows are LESS adversely affected by long intervals between milkings because lower-producing cows have less secretory tissue in the udder. Consequently, the increase in udder pressure (between milkings) per unit milk is greater in younger and lower-producing cows. Short milking intervals are thus important for young cows. Incomplete milking (e.g. leaving 2 kg milk in the udder at each milking) causes a permanent depression in milk production.

Frequency of milking

The rate at which milk is secreted is important in determining how much extra milk will be produced when cows are milked more often than twice daily. On the average, production will be increased by 20% when milking 3 times per day as compared with twice daily. Milking 4 times per day yields a further 5 to 10% increase.

The additional yield occurs because :-

- The milk secreting cells operate at full capacity for longer,
- Of better feeding and management often associated with 3 times per day milking.

Decreasing the number of milkings per day causes a marked reduction in milk yield. Skipping one milking a week reduces total yield by 5 to 10%, while milking once daily cuts the yield by half in first calvers, and by 40% in older cows.

Milk Removal

A vital part of lactation is of course the process by which the milk is removed. This includes :-

- Passive withdrawal from the cisterns and ducts of the udder
- Ejection of milk from the alveolar lumen.

To assist in expulsion of milk from the udder, the myoepithelial cells are arranged in two ways.

- In a starlike (stellate) fashion around the alveoli to achieve a squeezing action and to collapse it. This is called implosion.
- In a longitudinal (lengthwise) arrangement along the small ducts to effect a shortening and widening of these small tubes. This aids the expulsion of the milk from the collapsed alveolus by increasing the size of the passage through which the milk must pass.

These myoepithelial cells have no nerve connection and are controlled by hormones.

In contrast to the myoepithelial cells within the udder, smooth muscle occurs around :

- Blood vessels
- Small and large ducts
- Sinuses

This smooth muscle is controlled by nerves which pass from the spinal column through the glandular tissues terminating at the muscles of the teat sphincter.

Passive withdrawal

In this phase of milking the active co-operation of the cow is NOT required. Nerve endings in the skin of the udder and of the teat are stimulated by washing/wiping of the udder, the slipping-on of the teat cups and by pressure of the teat liner on the walls of the teat. The teat sphincter at the tip of the teat then opens and milk begins to pass out into the milk tube. This phase, which is activated by the nerves, begins about 5 to 10 seconds after the udder and teats have been stimulated. The nervous reflex brings about a squeezing of the ducts by the smooth muscle and the milk within the ducts is pushed into the udder cisterns. Since the dairy cow has a relatively large cistern, 40 to 50% of the total milk yield may be obtained during this passive withdrawal phase.

Milk ejection reflex

Whilst passive withdrawal of milk in the cisterns is taking place, nerve impulses pass from the teat to the brain in response to stimulation from attaching the milking machine (Figure 10). Other messages picked up via the cow's nose and eyes are also transmitted to the brain by nerves. This information causes the brain to release the hormone oxytocin into the bloodstream. Eventually the oxytocin reaches the udder where it causes the myoepithelial cells around the alveoli to contract. Upon contraction of the myoepithelial cells, the alveoli collapse and milk is squeezed out into the small ducts. These ducts shorten and widen, and the milk rushes through into the gland cistern. Only alveolar and ductal milk is expelled by the action of oxytocin on the myoepithelium. There is no contraction of large ducts or cisterns.

The milk ejection reflex

This milk ejection reflex is pre-programmed (i.e. conditioned) by the brain. In other words, the brain interprets certain stimuli to mean that milking is about to commence. Accordingly, when the brain receives the correct message it automatically proceeds to trigger the release of oxytocin. The conditioned stimuli for milk ejection include rattling of the milk buckets, washing of udders, feeding of concentrate, approach of the milker, and application of the milking machine or massage of the udder prior to application of the teat cups.

An effective milk ejection reflex is needed to achieve fast, efficient milking. It is vital that the cow is able to recognize the signal from which she can anticipate milking to commence within a relatively fixed time, in other words, the signal must be clear and should always occur at a set time before the machine starts milking.

The milk ejection reflex usually takes place 20 to 40 seconds after the initial stimulation (e.g. udder wiping) and lasts for only about 6 minutes. It is most important that the milk is removed while this reflex is operating. The udder should be handled no earlier than 30 seconds before teat cup application, otherwise the beneficial effects of good milk ejection will be lost. In many situations, application of the teat cups themselves is THE stimulation for milk ejection. There will, of course, be 30 to 60 seconds lag between applying the machine and contraction of the alveoli. However, passive withdrawal will be occurring during this interval. This kind of lag may be particularly beneficial when milking high-producing cows since the reduction of pressure within the cisterns (by passive withdrawal) will aid flow from the alveoli. Once the effect of oxytocin wears off, milk that has not been removed will flow back from the udder sinuses into the ducts and eventually into the alveoli. It can then be removed only after oxytocin has again been released, which will probably occur only at the next milking. The extra pressure within the udder will reduce the synthesis of milk after the incomplete milking, and if this situation occurs repeatedly the cow will reduce her yield accordingly. In dairy cows the milk ejection reflex is probably essential for 50 to 60% of the total yield at a milking.

Milk ejection - the contraction of the myoepithelial cells surrounding the alveolus forces milk out of the lumen into the ducts.

Inhibition of milk let-down

The milk let-down reflex of cows is very easily retarded or prevented by external stimuli which disrupt the NORMAL routine for milking. It is often said that a cow "holds up" her milk. This is NOT possible since the cow has no voluntary control over the let-down process, and she cannot thus put the process into REVERSE. However, let-down can be prevented, or terminated, by nervous stimuli such as rough treatment of the cow, loud, unfamiliar noise, pain and irritation. Such stimuli cause the brain to release adrenalin. This hormone works against oxytocin by:

- Blocking oxytocin release from the brain
- Constricting blood vessels and preventing oxytocin from reaching the udder
- Directly counteracting the effect of oxytocin on the contraction of myoepithelial cells.

Pattern of milk flow from the udder

If adrenalin release occurs before the milk-ejection stimulus, the ejection will be virtually completely blocked. When adrenalin is released after milk ejection has commenced, it will result in large amounts of milk being retained in the udder with associated negative effects on milk yield.

Cows handled gently and milked carefully at regular intervals seldom suffer from this problem. Milking should be a pleasant experience for the cow in order to capitalise fully on milk ejection.

Milking

MILKING - Lactation is the continuous secretion and storage of milk in the udder. The milk ejection or 'let-down' reflex effect is short term, inhibited by pain or fear but stimulated by good husbandry practices. Even so, at least 10% of secreted milk will be retained in the udder as residual milk. Removal of milk is achieved when external forces such as suckling or milking open the teat duct at the teat end.

MILKING INTERVALS affect the amount of residual milk carryover between milkings. Equal intervals of 12 hours give highest lactation yields but the effect of unequal intervals is small up to 16 and 8 hours and can be minimised if the highest yielders are milked first in the morning and last in the afternoon.

Dairy farmers, with varying levels of skill, knowledge and resources, maximise returns from milk production by influencing lactation through selective breeding and control of reproduction, nutrition, disease and general management. The methods of milking have a particularly important effect because a cow cannot secrete over a period more milk than is removed by milking. Thus, maximising milk removal in ways which are economic will take fullest advantage of secretion potential.

Lactation

Lactation includes both milk secretion and storage in alveolar cells and ducts within the mammary gland, followed by milk ejection (let-down) and milk removal. Milk secretion is continuous and usually at a constant rate for at least 12 hours resulting in a gradual increase in internal udder pressure. Milk ejection is a neuro-hormonal reflex initiated by various stimuli at milking time. These stimuli, which reflect good husbandry practices, are either natural (inborn) or conditioned (learned by experience), including, for example, feeding and udder preparation. They cause the alveoli and small milk ducts to contract forcing milk towards the udder sinus. Once this has happened most, but not all, of the milk can be removed when external forces such as suckling or milking open the streak canal (teat duct) at the teat end, but at least 10% will be retained in the udder as residual milk.

Milking intervals

The 10%–20% of the secreted milk which is not expressed from the secretary tissue and is retained in the udder when milking is completed is called 'residual milk' and has a much higher fat content than even the end-of-milking strippings. The quantity of residual milk is proportional to total yield, so that with unequal milking intervals there is a larger net carryover of milk fat from the longer night-time to the shorter daytime interval. This accounts for the apparent faster secretion rate and higher fat content of afternoon milk production. Milk yields, particularly from higher yielding cows are usually greater when milking intervals are 12 hourly. The effect of uneven intervals is not large up to 16 and 8 hours, and can be minimised by milking the higher yielders first in the morning and last in the afternoon.

MILKING FREQUENCY affects total daily production. Milking three times instead of twice daily will raise milk yields by an average 10%–15% but up to 10% of this increase will be required to cover extra costs. Chemical composition of milk is unaffected.

INCOMPLETE MILKING either from excessive amounts of residual milk or end-of-milking strippings can be avoided by effective milk ejection stimulation and by efficient fast milking by hand or machine.

Milking frequency

As a general rule, herd lactation yields will rise as the frequency of milking is increased. On average, the rise in milk yields will be between 10% and 15%, the largest increases occuring amongst heifers. The chemical composition of the milk (fat and solids-not-fat) will be unaffected. Recent commercial data from developed dairying areas also reveal that, on average, up to 10% increase in yield is required to cover the extra costs of milking three times daily. The full benefit of the increased frequency is obtained by milking three times daily throughout lactation rather than reverting to twice daily when milk yields begin to fall. The reasons for the increase in lactation yields are inconclusive; the most likely being the more frequent removal of secretion inhibiting substances which begin the drying-off process.

Incomplete milking

There are two forms of incomplete milking. One is that excessive amounts of residual milk are retained in the udder because of inadequate milk ejection stimuli or the inhibitory effects of adrenalin secreted by cows becoming frightened and upset during milking, or even by slow milk removal. The other form is that some of the available milk is left in the udder when milking ceases, ie., the so-called 'strippings'. The modern milking machine is designed to remove 95% of available milk without recourse to additional cluster weight or manual assistance. Hand stripping, particularly with the finger and thumb should be avoided. The amounts of strippings are likely to be small even in relation to normal levels of residual milk and if not removed are unlikely to affect significantly either the lactation yield or quality of milk.

MILKING ROUTINES are a reflection of good stockmanship. Cows are creatures of habit; avoid any circumstances which upset or frighten them and so inhibit milk ejection.

Residual milk amounts will be inversely proportional to the strength of the stimuli signals. Develop a regular and repetitive milking routine. Make changes gently and carefully. Milk quickly and quietly in a stress-free environment.

Milking routines

The aim of an efficient and effective milking routine is to leave the least amount of residual milk in the udder. This, in itself, is a measure of good stockmanship. Milk ejection can be stimulated manually by a series of activities carried out by the person doing the milking. The amount of residual milk is inversely proportional to the strength of the conditioned stimuli signals, which are developed into a regular, repetitive milking routine, including such activities as feeding and udder preparation. The stimulation response is transitory, the maximum effect declining within a few minutes of milk ejection occuring. Therefore delayed milking will reduce the amount of milk removed. The internal pressure of milk within the udder peaks between one and two minutes after milk ejection and therefore milking should be completed as soon as possible after this occurs. Cows are creatures of habit and consequently changes to the routine should be made gently and quietly. It is important to avoid any circumstances which upset or frighten them causing the release of adrenalin which adversely affects the circulatory and musculatory systems, thus restricting effective milk ejection and prolonging the duration of milking. The response of cows and those milking them to a pleasant and stress free environment will be measured in terms of production levels.

Because residual milk and strippings have fat percentages that normally exceed 10%, incomplete or slow milking can reduce markedly the fat content of the milk at any particular milking. However, it is important to recognise that milk fat retained or left in the udder is not lost but will be obtained at succeeding milkings. In fact, although management factors (eg. varying milking intervals and milking frequency) may alter the fat content of milk at one milking, the average fat content over a period of time will be unaffected. On average, the fat content of milk obtained must be the same as that secreted into the udder. The concentrations of protein, lactose and other solids-not-fat are unaffected by changes in milking management either at one or more milkings.

HAND MILK with clean, dry hands. Use the full hand and avoid finger and thumb stripping.

MACHINE MILKING will create a pleasant sensation for the cows if the machine is kept clean, maintained properly and operated according to the manufacturer's instructions. Attach and remove clusters carefully to avoid vacuum fluctuations which cause mastitis. Readjust slipping teatcups and replace fallen clusters immediately.

Milking methods

Hand milking should always be done using clean, dry hands. Preferably, milk with the full hand and avoid end-of-milking stripping with the finger and thumb. Rear quarters should be milked first as they contain most milk and the milking bucket hooded to reduce contamination from dust and udder hairs.

Methods of machine milking are designed to create a pleasant milking sensation for the cows and to avoid any possible hazard to udder health. It is most important that milking is done with a well designed, carefully cleaned and properly maintained machine which is operated strictly according to the manufacturer's instructions.

A skilled operator pays particular attention to careful cluster attachment and removal from the udder. During cluster attachment it is essential to ensure that the vacuum cut-off arrangements to the clawpiece are effective so that excessive volumes of air do not enter and cause vacuum fluctuations in the main vacuum pipeline system, as this could increase mastitis incidence. Attach each teatcup carefully starting with the two furthest from the operator. The clusters are removed as soon as milk flow ceases, avoiding excessive air entry through the teatcups by cutting off the vacuum supply before gently but firmly pulling the teatcups from the udder. During milking, any teatcups which slip from the teats should be readjusted immediately and any clusters which fall to the floor should be cleaned and reattached without dela

5

Cleaning and Disinfection in a Dairy

The hygienic quality of milk affects the health of the consumer and the quality of the end product, and is therefore commercially important. The dairy farmer has the responsibility of producing milk under clean and hygienic conditions, employing approved techniques and procedures to clean and disinfect the milking equipment and milking barn.

The safety of raw milk for the consumer and the ability of raw milk to retain its quality during storage directly relate to the bacterial content of the milk. In most countries, bacterial content is one of the factors considered in the milk payment scheme.

When producing raw milk of good bacteriological quality, cleanliness and disinfection of the udder and the milking equipment are the most important requirements, even more so than cooling the milk. Cleaning and disinfection are complementary processes; neither process alone will achieve the desired result, which is to leave the surface of milking utensils as free as possible from milk residues and bacteria.

The bacteriological quality of milk is determined by various aspects of dairy management and husbandry practices. Milk is sterile when secreted by an udder gland which is free from infection. Contamination with microorganisms occurs during and after milking. As the average number of dairy cows per herd has increased in recent years, greater numbers of animals have tended to be concentrated in each production facility, thus creating a greater risk of exposure to pathogens for the individual animal, due to increased bacterial contamination of the premises. High levels of moisture in the dairy farm environment allow bacteria to grow and multiply, and increase the possibility of bacteria reaching the teat apex, leading to a greater risk of bacterial penetration into the mammary gland.

Commercial dairy operations in most countries have increased in size and complexity. Automation is replacing manual labour in the barn. With automated removal of the milking unit, the operator is no longer required to return to the cow at the end of milking, and therefore post-milking teat disinfection is frequently inadequate or is abandoned altogether. Workers responsible for a large number of cows and milking units have little opportunity for observation of individual cows, and thus mild cases of udder diseases may pass unnoticed and untreated. The availability of automated systems has prompted many dairy producers to introduce thrice-daily milking, and continuous genetic improvements have increased milk yield, thus creating an environment conducive to a higher level of udder infection and a heavier charge of contaminating microorganisms. Hygiene plays a primary role in reducing the burden of the many pathogens present in the dairy farm environment.

This article considers the sources of bacterial contamination of raw milk, and the recommended cleaning and sanitation practices necessary for the production of milk of good microbiological quality. Milk quality is determined by the following factors:

(a) Udder diseases or 'internal' milk contamination
(b) Contamination of milk from the udder and teat skin
(c) Contamination of milk from milking equipment and the milking barn
(d) Multiplication of bacteria in milk during storage in the farm and transport to the processing plant.

UDDER DISEASES AND MILK CONTAMINATION

Mastitis is the most widespread infectious disease affecting dairy cattle. The chronic, sub-clinical form of mastitis may affect cows throughout their productive life, while the clinical form usually attacks high-producing cows in the first months after parturition. Both forms of the disease cause deleterious changes in milk quality and composition. Infected glands cause fluctuating numbers of microorganisms to be excreted in milk, the highest numbers being found in foremilk.

Bovine mastitis is a multifactorial disease complex resulting from the interaction of the cow, the environment and the microorganisms pathogenic for the udder. Major losses from mastitis include those due to decreases in milk production, the cost of replacement heifers, loss of unsaleable milk following the administration of antibiotics, the cost of drugs, veterinary treatment, extra labour, and the loss of genetic material from the herd.

Sub-clinical mastitis is caused by udder pathogens which multiply in the mammary gland and are essentially contagious, being spread between cows principally at milking time in the parlour, due to contact with the hands of the milker, udder cloths and the rubber liner of the milking unit. The main pathogenic organisms involved are staphylococci, Streptococcus agalactiae and Streptococcus dysgalactiae.

The bacteria causing clinical mastitis multiply principally in faeces and bedding, and in the water supply. Thus, udder contamination principally occurs via the environment outside the parlour, the main 'environmental' organisms being Escherichia coli and coliform enteric bacteria, Streptococcus uberis and Actinomyces pyogenes.

Control of sub-clinical contagious mastitis must concentrate on the cows, the milking parlour or the barn, and may be achieved by teat disinfection (post-milking teat dipping) and treatment, preferably during the 'dry' period. On the other hand, control of environmental clinical mastitis-causing pathogens must concentrate on general cleanliness and dryness, ventilation, effective disposal of discharges, fly control and adequate sanitation of the milking machines. The hygienic preparation of the udder before milking is essential for the control of both forms of mastitis, although such treatment is more effective in the prevention of the clinical form (27, 28). The infecting bacteria lead to mammary gland response, including stimulation of the migration into the udder of leukocytes or white blood cells which ingest and kill invading bacteria, releasing compounds that cause inflammation, tissue damage and alteration of the secretory functions of the gland. Mastitis alters milk composition: lactose and fat levels fall, and there are major changes in milk proteins (decrease in casein). Sodium and chloride ion concentrations in the milk rise and the level of potassium falls. This causes increased electrical conductivity, a factor which is used for detection of mastitis; other methods include counting the somatic cells (both leukocytes and desquamative epithelial cells) in milk. This last marker for the detection of mastitis (mainly sub-clinical), together with the total bacterial count, is currently employed in many countries for calculations in milk quality payment systems (Table I). Clearance of udder infection in dairy cows by treatment, segregation and culling will not be considered here. This discussion will concentrate on the techniques of prevention and hygienic milking recommended for diminishing the rate of new infection with contagious sub-clinical mastitis and thus improving the bacteriological quality of raw milk.

Hygienic procedures in controlling mastitis

For practical reasons, milking parlour sanitation can be separated into three main phases: pre-milking sanitation, cleansing of the milking unit after each cow, and teat disinfection after milking.

Teat disinfection after milking will be dealt with in the next section, while the other two phases will be covered in the sections ‘Contamination of milk from the udder and teat skin’ and ‘Contamination of milk from milking equipment’.

Teat disinfection after milking

The value of post-milking teat disinfection is now widely accepted. Farmers who consider this procedure too laborious and expensive should be reminded that it takes an average of 4-6 seconds per cow to apply, uses approximately 7.5 1 of disinfectant per cow per year for twice-daily milking (11.25 1/day for thrice-daily milking) and costs approximately US$23 and US$35 annually for twice- and thrice-daily milking, respectively.

Teat disinfection aids in reducing contagious mastitis infections caused bystaphylococci, Streptococcus agalactiae and the Mycoplasma species. There is good evidence that, when practised continuously, teat disinfection after milking reduces new udder infections due to these organisms by 50% or more, although this practice appears to offer very little protection against environmental pathogens, such as coliforms.

Techniques

The most widely-used methods of teat disinfection are described below.

Manual teat dipping

Manual teat dipping provides good coverage of the teat skin and is the most popular method of teat disinfection. The disinfectant solutions are applied from an open cup or a closed, non-spill cup (which reduces wastage and requires less refilling during milking). When sufficient clean disinfectant solution is maintained within the dip cup, the teats of approximately 25 cows may be dipped after milking without refilling the cup with fresh solution. The solution remaining in the cups should be disposed of after the completion of milking, and the cups should be washed clean and maintained dry until the next milking. Some milkers object to using disinfectants, and occasionally suffer allergic reactions from disinfectants splashing onto their hands.

Manual teat spraying

Manual teat spraying involves the use of a pressurised container fitted with a suitable nozzle, best positioned at the curved end of an elongated spraying tube. Milkers who prefer spraying claim that less disinfectant is used and that the solution is always cleaner and of more uniform concentration than that used for dipping. However, when spraying is performed quickly, teats may be missed; in udders with abnormally-positioned teats, a teat may be shielded by a leg. The risk to operators of continuously inhaling the chemical droplets has motivated the wider use of the dipping technique.

Automatic teat disinfection

Devices for automatic teat disinfection after milking are usually cow-activated and are positioned at the exit from the parlour. Most devices are experimental and only a few are sold commercially. When built and operated correctly, these devices can achieve good coverage of the teat ends and teat sides. As there is no need for workers to be close at hand during operation, no inhalation of chemical droplets occurs. A disadvantage of automatic devices is that they use twice as much disinfectant as manual spraying.

Teat disinfectants

Among the disinfectant products most widely used for teat disinfection are iodophors, chlorine-releasing compounds and Chlorhexidine. Most commercial preparations add emollients or dermatological excipients, such as glycerine or lanolin, to improve teat skin tolerance under continuous use.

Iodophors

Iodophors are the most widely-used teat disinfectants in most countries. These include certain surface-active agents (usually a non-ionic surfactant) which solubilise iodine and form compounds possessing the germicidal activity of iodine without the toxic and irritant properties. Iodophors are microbicidal. When iodophors are diluted with water, the iodine micelles aggregates are dispersed and most of the iodine is slowly liberated. The microbicidal activity encompasses bacteria, bacterial spores, fungi and their spores and some viruses. Iodophors remain potent in the presence of organic matter, provided that the pH does not rise above 4. The activity of iodophors decreases markedly when diluted in water with a high level of alkaline hardness. Iodophors are formulated with phosphoric acid to provide a pH of approximately 3 for optimal antimicrobial activity. Povidone-iodine (a water soluble complex of iodine and polyvinylpyrrolidone) is a general antiseptic used in some countries for teat disinfection.

The recommended concentration (free iodine) of commercial iodophors for teat disinfection in most countries is 0.5% (5,000 ppm). In some regions, higher concentrations (1.0%) are used, but a tendency towards using lower disinfectant concentrations during milking has recently been observed in some countries. This policy is aimed at reducing the risk of milk contamination with chemical residues, thus ensuring a higher-quality and more wholesome product for consumers. Published evidence is available on the efficacy of iodophors for teat disinfection at concentrations of 0.25% or lower. With these lower concentrations, more frequent refilling of the dip cup is required: each refill may be used for up to 15 cows instead of 25. The staining properties of iodine are present, although diminished, in iodophor preparations; such staining is useful as it shows whether the cows have been correctly treated.

CHLORINE-RELEASING COMPOUNDS

The most commonly-used chlorine-releasing compound is sodium hypochlorite, which has a wide antibacterial spectrum, although it is not effective against mycobacteria. The activity of sodium hypochlorite is markedly affected by organic matter and light. Chloramine compounds have also been used as teat disinfectants. Both types of disinfectants are more stable in alkaline solutions at concentrations varying from 0.1% to 4% (1,000-40,000 ppm) without emollients, and with a pH between 8.5 and 11. These compounds are active against staphylococci and Streptococcus agalactiae and are usually cheaper than other disinfectants. The limited use of chlorine-releasing compounds compared with the other groups is explained by their pungent odour, and their capacity to induce decolouration of fabrics and some irritation of teat skin. Isocyanure is a less irritating chlorine-releasing compound used in some countries, which is usually supplied in the form of effervescent tablets and is recommended for use at concentrations of 0.5% (5,000 ppm).

Chlorhexidine

Chlorhexidine is a member of the biguanide group of disinfectants; it has bactericidal properties against Gram-positive and Gram-negative bacteria and is antifungal, being more active at alkaline pH. As this product is colourless, some commercial preparations include the addition of approved artificial colourants, which leave a mark on the teat skin similar to that left by iodophors.

Chlorhexidine has a longer residual activity on teat skin than other groups of disinfectants. This property is important, due to the increased contact time for antibacterial activity on the resident skin bacterial population and the preventive activity against skin contamination with environmental organisms during the period after milking (when the teat canal is not yet closed and the danger of bacterial penetration into the gland is increased). Organic matter does not greatly affect the activity of Chlorhexidine. Recommended concentrations for teat disinfection are usually higher than with iodophors, i.e. approximately 0.5% (5,000 ppm). As the solubility of Chlorhexidine depends on water quality, the daily preparation of solutions is recommended in regions with alkaline hardness in the water supply.

Other disinfectants with limited use for post-milking teat disinfection are quaternary ammonium compounds (used in Canada, France and the United Kingdom), glutaraldehyde (Switzerland, France and the United Kingdom), linear dodecylbenzenesulfonic acid (LDBSA) (Canada, the United States of America [USA], Denmark, Finland, Spain, France, Belgium and the United Kingdom), chlorous acid and chlorine dioxide (Canada, the USA and France).

Disinfectants should be formally approved and registered. In most countries, registrations are performed by the Ministry of Agriculture, and approval and registration must be clearly marked on the product packaging, together with complete instructions on dilution and application, and warnings of potential risks to animal and human health and of possible residues in milk.

In view of the need to protect the teat duct orifice, which is usually open to some degree for a short period after milking, disinfection of teats immediately after milking should be recommended.

CONTAMINATION OF MILK FROM THE UDDER AND TEAT SKIN

Sanitising the udders of cows prior to milking reduces the opportunity for environmental organisms to gain entrance into the raw milk and reduces exposure of the udder to these organisms.

The many types of microorganisms which colonise the external surfaces of the udder and teats include faecal, soil and water coliforms (E. coli, Enterobacter aerogenes, Klebsiella spp.), streptococci (mainly S. uberis), staphylococci, coryneform organisms, bacilli, yeasts and fungi (18). Continuous post-milking teat disinfection greatly reduces the population of organisms on the teat skin. However, such procedures have almost no effect on bacterial populations obtained by cows lying down in an infected environment, or contamination due to contact with the hands of milkers, washing water or cloths. Wet skin is known to shed more bacterial colonies into milk than dry skin. It has been demonstrated that different methods of pre-milking udder preparation significantly influence milk quality. The lowest bacterial counts in milk were observed when teats were cleaned with water followed by thorough drying with individual paper towels, or when a disinfectant was applied to teats followed by drying (12). Pre-milking teat disinfection by teat dipping in association with good udder preparation reduced the rate of intramammary infections due to environmental pathogens by 50% (compared with good udder preparation alone) (24). A common practice is to wash but not dry the teats; this has a deleterious effect on the bacteriological quality of milk, increasing the risk of coliform and S. uberis mastitis, as these bacteria are often present on the teats before milking and might run down with the water onto the teat apex in the course of milking.

A basic approach to recommended preparation of cows for milking in the milking parlour should include the following elements:

(a) Identify cows (udder quarters) secreting abnormal milk and discard this milk; it is not fit for human consumption and should not be mixed with the milk of healthy cows for further processing. In most countries, this point is enforced by official regulations. Foremilk examination is the technique used to identify abnormal milk before applying the milking unit. Although milk filtration removes larger particles of dirt and sediment, it cannot remove bacteria, and thus a filter does not provide an alternative to good management and husbandry practices, if good raw milk quality is the required goal.

(b) Clean and disinfect the teat skin, thus reducing the bacteriological contamination of raw milk. The efficacy of teat washing is increased if an approved disinfectant is added to the wash water. dditionally, washing teats one minute before milking provides a stimulus for the milk ejection reflex.

Inevitably, handling and washing the teats of cows in the milking parlour transfers pathogenic organisms between teats and between cows. To minimise this transfer, it is necessary that a disinfectant is included in the washing solution, that an individual clean cloth or paper towel is used for each cow, and that the hands of the milker are washed and dried before preparing each cow for milking. Wearing rubber gloves is advisable but this practice is not widely accepted by milkers, particularly in hot climates.

The disinfectant employed is usually either chlorine (in a concentration ranging from 100 to 300 ppm) or an iodophor (in a concentration between 25 and 75 ppm). Cloths or towelettes impregnated with disinfectant were found to be effective for pre-milking teat preparation (1), and should be recommended for individual cow use (one towelette per cow).

(c) A complement of pre-milking udder and teat preparation, developed in the USA and indicated mainly to reduce the incidence of environmental mastitis, is pre-milking teat disinfection by dipping or spraying with chlorine-based disinfectants or iodophors usually at lower concentrations than for post-milking disinfection. Chlorine compounds are used for pre-milking teat disinfection at 20,000 ppm (in the United Kingdom) and iodophors are used at between 1,000 and 2,500 ppm. In some countries (Spain, the USA and Canada), Chlorhexidine is used at 1,100-5,000 ppm. LDBSA is used at concentrations of 19,400 ppm (in the USA and Denmark), and chlorine dioxide and hypochloric acid at 30,000 ppm (also in the USA and Denmark). Nisin, a proteinaceous antimicrobial bacteriocin, is used in France. In pre-milking teat disinfection, the formulation must be applied to the whole teat surface, which has been previously cleaned. After a minimum of 10 seconds, the teats must be wiped dry with a paper towel to remove excess disinfectant. It should be stressed that pre-milking dipping or spraying is not a substitute for cleaning and washing teats in the usual way; these procedures are simply additions to a good pre-milking hygiene routine and do not represent alternative methods of udder preparation.

The numbers of bacteria shed in the milk from an infected udder may exceed 109/ml, particularly in cases of clinical mastitis. Even a single case of mastitis can significantly increase the bulk milk bacterial count in a medium-sized herd. Dirty teats are the main source of visible extraneous matter or sediment content, and may lead to bacterial counts approaching 100,000/ml in milk. Heavily-soiled, unwashed teats may produce an unacceptable sediment in milk. However, when cows are on pasture in dry weather, the bacterial count of milk from unwashed teats may be less than 100/ml.

CONTAMINATION OF MILK FROM MILKING EQUIPMENT

Surfaces of milking and cooling equipment which come into contact with milk are the main source of milk contamination, and are frequently the principal cause of consistently high bacterial counts. Relatively simple and inexpensive cleaning and disinfection procedures can effectively diminish bacterial contamination from this source in milk. However, faults arise which may result in a build-up of potentially pathogenic bacteria, mainly pseudomonads, coliforms and enterococci, on the milking equipment surfaces. Common among such faults are inadequate water temperatures, over-dilution of detergent sanitisers, failure to ensure adequate contact time with the soiled surfaces, and rubber inflations (liners of the milking unit) and tubes in poor condition. Although rubber inflations are not contaminated for the first cow being milked with the unit, they potentially transfer pathogenic and saprophytic bacteria originating from within the udder, from the udder and teat skin, and from faeces to cows subsequently milked with the same unit.

It is virtually impossible to remove all milk deposits and residues from the milkcontact surfaces of milking equipment. Bacteria multiply on these surfaces in the interval between milkings, especially in regions with a hot or moderate climate. High numbers can be present on apparently clean equipment. It is essential to recommend the use of milking equipment with smooth milk-contact surfaces, with a minimal number of joints, dead ends and crevices, the use of a water supply of drinking quality, detergents to remove milk residues and sediment, and disinfectants to eliminate bacteria. Due to the complexity of milking machines, cleaning and disinfection of milk-contact surfaces may not be fully effective, so that milk residues and bacteria are not completely removed from the equipment and thus accumulate daily. The extent of bacterial contamination on milk-contact surfaces of equipment is usually determined by rinsing with a sterilised liquid. This liquid is then subjected to a bacterial count in the same way that milk is tested, and any bacterial species present can be determined.

Methods of cleaning and disinfecting the milking equipment will be described below, with reference to the various milking systems generally employed. Selection, storage and handling of disinfectants and detergents are also discussed.

Manual cleaning

Washing by hand is appropriate for bucket milking machines and clusters, and for ancillary equipment (e.g. receiving pans, strainers and coolers) used when milk is put into churns. Manual cleaning may also be used for smaller bulk tanks. The daily cleaning routine consists of three stages: a rinse with cold or tepid water (38°C), a warm detergent wash and a final rinse with clean water.

The cold or tepid water rinses remove residues of milk which would otherwise partially inactivate the disinfectant in the next stage. Tepid water is particularly beneficial for rinsing the clusters, as this is more effective than cold water in removing fat and milk residues. After milking, the outside of the milking unit is cleaned. Each unit should be rinsed by connecting the vacuum tube to a vacuum tap and drawing clean water through the teat cups. If the clusters are not cleaned immediately, they should be left immersed in water. The next stage, the warm detergent-disinfectant wash, is the most important of the three. Unless the solution reaches all milk-contact surfaces, milk residues may remain which will protect bacteria from the disinfectant. Care must be taken to avoid air-locks in the clusters. A volume of 45 1 of detergent-disinfectant solution is sufficient for up to three milking units and ancillary equipment. Clusters should be washed first by full immersion for two minutes; teat cups, milk tubes and claws are scrubbed, and the cluster is re-assembled and transferred to the rinsing trough. For the final clean water rinse, the addition of 50 ppm of hypochlorite significantly improves results. The clusters should be rinsed and hung up to drain.

In case of deposits on the milk-contact surfaces or high bacteriological counts, extra steps are required, as detailed below.

De-scaling with acid

Phosphoric acid is used to remove milk stone (milk and hard water residues). After cleaning, the metal components of the milking unit are dismounted and soaked in the de-scaling solution, in accordance with the instructions of the manufacturer. All parts should be brushed with detergent-disinfectant solution and finally rinsed. Eye protection and rubber gloves should be worn when handling the acid.

Heat treatment

Should detergent-disinfectant solution be unavailable, heat treatment is necessary. Approximately 9 1 of hot water (85°C) should be rinsed through the clusters. A domestic boiler is a convenient source of hot water. After treatment, the equipment is hung up to dry.

Wet storage of clusters

Clusters are suspended in a rack in such a way that they can be filled with a suitable detergent-disinfectant solution between milkings.

In the case of milk rejection due to high bacterial counts, immediate attention should be given to the following points:

- Old and worn rubber parts should be replaced, together with metal or plastic equipment which is rusty, corroded or has open seams.
- Metal components should be de-scaled and rubberware soaked in hot detergentdisinfectant solution.
- Daily cleaning and disinfectant methods should be checked and any faults corrected.

In-Place Cleaning

The predominant method of in-place cleaning for milking pipelines and recorder machines is circulation or re-circulation cleaning in barns and parlours. Less common methods, such as the use of acidified boiling water, will also be mentioned below.

Circulation Cleaning

Circulation cleaning is a three-stage process consisting of a pre-rinse with water, a recirculated hot wash with detergent-disinfectant solutions and a final cold water rinse. For cleaning particular installations, most manufacturers issue instructions in conjunction with a diagram of the milking installation. The efficiency of circulation cleaning depends on the temperature of the water used for the detergent-disinfectant wash, the optimum initial temperature being at least 85°C. Parlours with large-bore pipeline systems have air-injectors to develop turbulence of the cleaning solution and thus improve surface contact and disinfection efficiency. In parlours with recorder jars, spreader plates and valves are fitted at the top of the jar to improve contact with the internal surfaces, thus ensuring sufficient solution flow through the clusters.

Procedures for circulation cleaning are usually indicated by the milking machine manufacturer. Although some differences exist between the various types of milking installations, the generally-accepted procedures are as follows:

1. After milking, rinse the machine thoroughly with warm water, and brush the clusters to remove external dirt.
2. Attach the jetters to the cluster.
3. Check the water temperature (85°C).

4. Drain the milk from the receiver and milk pump.
5. Remove the filter sock (interior) and clean the filter as recommended by the manufacturer.
6. Connect the air pipeline directly to the water heater and set the three-way valve to the washing position so that the hot rinse water is drawn into the machine.
7. Set the releaser milk pump to run continuously and adjust the spreader on the receiver lid to the washing position.
8. Allow hot water to pass through the machine and discharge to waste until the temperature of the water leaving the machine exceeds 50°C.
9. Add approved detergent-disinfectant solution to the measured volume of hot water, in accordance with the specifications of the manufacturer. Set the three-way valve to draw solution from the wash through the installation and continue circulation for 5-10 min. No advantage is gained by prolonging circulation, as the temperature of the solution progressively falls.
10. Discharge the detergent-disinfectant solution by deflecting the delivery pipeline.
11. Run clean cold water into the machine. Sodium hypochlorite may be added at a concentration of 50 ppm to avoid risk of contamination from supply lines.
12. Switch off the releaser and vacuum pumps; drain and prepare the machine for milking.

If a hypochlorite rinse is given for two minutes immediately before milking, it is not necessary to add hypochlorite to the final rinse (step 11).

Some manufacturers provide equipment which automatically controls the circulation cleaning cycle and the dosing of detergent-disinfectant products. Regular and frequent inspection of the installation (performed after the cleaning process) is necessary to detect and correct defects, such as malfunctions in inflations, punctured short milk tubes, leaking rubber joints, flattening of any rubber tubes (which could result in incomplete circulation) and build-up of milk residues.

Acidified Boiling Water Cleaning

This method relies on heat for bacterial sanitation. For each cluster, 18 1 of boiling water is required, which enters the machine at a predetermined rate and is then discharged. In the first 2-3 min of cleaning, acid (nitric, sulphuric or citric) is added to the water to prevent hard-water salt deposits on internal surfaces. All components should reach 77°C and be kept at this temperature throughout the cleaning process, for approximately 5 min. All components of the milking installation should be checked to ensure ability to withstand the high temperatures and acidity. Extreme care should be taken when handling concentrated acids. Eye protection and rubber gloves should be worn when preparing diluted solutions. A recommended routine for acidified boiling water cleaning at the end of each milking (or alternatively once a day or once a week) is as follows:

1. Brush clean the outside of the clusters and jetters with a detergent-disinfectant solution and fit the teat cups into the jetters.
2. Check that the water temperature in the boiler is at least 96°C.
3. Remove the milk delivery pipe from the bulk tank.
4. Add the stock acid solution to the acid container.
5. Turn the three-way valve to the wash position and open the wheel valve controlling the flow from the water boiler.
6. Adjust the spreader on the receiver lid to the washing position
7. Set the releaser milk pump to run continuously, to discharge return water to waste.
8. When water flow ceases, stop the vacuum and releaser pumps and drain the machine.

The total flow time should be 5-6 min. All parts of the plant should reach 77°C after 2-3 min.

MULTIPLICATION OF BACTERIA IN MILK DURING STORAGE ON THE FARM

After the bulk milk tank has been emptied, the tank should be rinsed with cold water of drinking quality to remove residues and then cleaned with detergent-disinfectants, using cold or warm water (37°C). If cold water is used for cleaning, addition of iodophor solutions has proved to be effective and has the advantage of rendering film build-up visible.

Cleaning of the bulk milk tank can be performed manually or by mechanical methods.

Manual cleaning of the bulk milk tank

The general routine for manual cleaning is as follows:

1. Brush the dipstick and bulk tank plug with a detergent-disinfectant solution and seal the tank with the blanking plug.
2. Brush and scrub the inner surface of the tank with a detergent-disinfection solution.
3. Drain the solution to waste and replace the dipstick and plug before hosing with clean cold water.
4. Drain the tank and replace the plug. The tank is then ready to receive milk.

Variations to the manual procedure include brushing with special detergentdisinfectant powders which are applied to the wet inner surface of the tank. The powder is worked into a paste by brushing, and spread to cover all the surfaces; after a contact time of two minutes, the tank is rinsed by hosing with cold water. Alternatively, a handoperated spraying device which introduces iodophor into the stream of water from a cold water hose can be used, with a contact time of 10 min, before rinsing with cold water and draining the tank.

Mechanical cleaning of the bulk milk tank

The same steps described for manual cleaning apply to the mechanical cleaning of the bulk tank, which is performed after milk collection. This may be a totally automatic process or it may be restricted to the first water rinse and cleaning with the detergent disinfectant solution, leaving the farmer responsible for the final rinse.

The disinfectant is applied by means of a spraying device in the tank. Build-up of milk residues around the joint between the rubber plug and the metal shaft can be prevented by immersing these parts in boiling water for two minutes once a month. Frequent inspection of the interior of the tank and its fittings is necessary. The operation of the mechanical equipment should also be checked to ensure that the iodophor solution is evenly spread.

Monitoring and assessment of cleaning and disinfection includes inspection of the equipment and examination of the methods used. Bacteriological rinses and swabs from milking equipment are used to test the quality of the final product: raw milk.

Cleaning and disinfectant agents

The detergents and disinfectants used in the milking parlour or barn are usually registered or officially approved in different countries by regulatory institutions (usually the Ministry of Agriculture and Ministry of Health). Among member countries of the International Dairy Federation, official registration is required in Belgium, Canada, Denmark, Israel, Italy, Japan, the Netherlands, Spain, Sweden, Switzerland and the United Kingdom.

A clean water supply of drinking quality is required in the dairy to rinse equipment, prepare cleaning and disinfectant solutions, provide hot or near boiling water, and wash udders and teats.

Legally, all milking equipment must be cleaned and sanitised after each milking, and detergents, detergent-disinfectants and disinfectants are available for this purpose.

Cleaning is the most important part of the process, as stressed in the 'Introduction', and if milking equipment is not kept visually clean, chemical disinfectants are unlikely to be effective.

Detergents

The addition of suitable detergents to the water used for daily cleaning assists in the removal of milk residues by wetting the surface to be cleaned, displacing the milk soilage, emulsifying fat, dissolving protein, holding soilage in suspension, softening water and preventing sediments during rinsing. No single detergent has all the necessary properties, and therefore commercial detergents usually contain a blend of ingredients. The choice and proportion of the different ingredients are influenced by cost, adverse effects on the hands of operators or on equipment, and the method of application (manual or in-place cleaning systems).

Common ingredients in the detergent formulations are as follows:

- Inorganic alkalis, such as sodium carbonate (washing soda), sodium hydroxide (caustic soda), tri-sodium phosphate and sodium silicates.
- Surface-active agents, such as anionic and non-ionic wetting agents, which increase contact with surfaces.
- Sequestering agents (water softeners), such as polyphosphates and ethylenediaminetetraacetic acid, which chelate hard waters cations.
- Acids, such as phosphoric, nitric and sulphamic acids, which prevent the build-up of milk stone.

Detergents for washing by hand consist of milk alkalis with small proportions of wetting agents and water softeners. For mechanical cleaning, stronger alkalis are considered necessary (to compensate for lack of brushing), together with water softeners. Most commercial detergents are compatible with sodium hypochlorite at normal use concentrations. In dairies with a very hard water supply, extra polyphosphate is recommended. Acids, sometimes containing wetting agents, are used occasionally to prevent or remove milk stone.

Disinfectants

Disinfectants may be used alone, or in combination with detergents in 'detergentdisinfectant' formulations. The name 'detergent-disinfectant' - although universally accepted - may cause confusion as, in fact, a minimal amount of disinfectant is added and the actual disinfection efficiency of such formulations is low.

The main types of disinfectants used in dairies are listed below.

Sodium hypochlorite

Sodium hypochlorite is the most widely used disinfectant for milking installations, due to its efficacy as a bactericidal agent; it is also the cheapest disinfectant suitable for use in the dairy. Commercial formulations are supplied in concentrations of between 7% and 12% w/v of available chlorine, with up to 2% free sodium hydroxide. Sodium hypochlorite may be combined with a detergent to allow cleaning and disinfection to be performed in a single operation.

Chlorine-releasing compounds

The chlorine-releasing compounds commonly used in dairies include salts of di- and tri-chloroisocyanuric acid and dichlorodimethyl hydantoin.

Quaternary ammonium compounds

The foaming properties of quaternary ammonium compounds make these products suitable for manual cleaning and for use in processing plants, but not for circulation cleaning in combination with detergents.

Iodophors

Iodophors are complexes of iodine with surfactants, combining high iodine availability with low irritancy and odour, and formulated with acids to enhance the bactericidal properties. Iodophors are widely used for spray cleaning and disinfection of bulk tanks. Low-foam formulations are available for circulation cleaning.

STORAGE AND HANDLING OF CHEMICALS IN THE DAIRY

All chemicals should be kept out of the reach of children and all unauthorised persons. Registered chemical agents should be used before the expiry date indicated on the label. All concentrated chemicals should be handled with care and in accordance with the warnings on the label. Eye and hand protection is strongly recommended when handling caustic soda and concentrated acids.

Detergent-disinfectants lose strength when exposed to light and heat, and they should therefore be stored in a cool, dark and dry place. Different chemicals should not be mixed, as this may cause a dangerous reaction. The detergents and disinfectants used in milking parlours should not have a strong smell. The following should be clearly marked on the containers of an approved disinfectant:

- A notice that they contain an approved disinfectant
- The specific purpose for which the disinfectant is approved and the dilution rate for recommended use in the milking parlour
- Warning notices to the user
- Expiry date and batch production number
- Complete identification of the manufacturer, importer or supplier of the product.

Cleaning and Sanitizing of Containers and Equipment

"The product contact surfaces of all multi-use containers, utensils, and equipment used in the transportation, processing, handling, and storage of milk or milk products shall be effectively cleaned and shall be sanitized before each use. Provided, that piping, equipment and containers used to process, conduct or package aseptically processed milk and milk products beyond the final heat treatment process shall be sterilized before any aseptically processed milk or milk product is packaged and shall be re-sterilized whenever any unsterile product has contaminated it."

Cleaning Operations

This item requires that all milk contact surfaces be effectively cleaned and sanitized before each use. The only exception to this is the Ordinance provision that milk storage tanks be emptied and cleaned at least every 72 hours and raw milk and heat treated milk storage tanks use to store products longer than 24 hours and all raw milk silo tanks be equipped with a 7-day temperature recording device.

This recorder shall have a scale span of not less than 50 F, be accurate to plus or minus 2 F, include the normal storage temperatures plus and minus 5 F, with 2

F minimum scale divisions not less than 0.040 inch apart and time scale divisions of not more than 1 hour. The recording chart of these devices must be capable of recording temperatures up to 180 F.

Computer generated temperature recorders which provide a printout which is readily discernible and meets the intent of the Ordinance are acceptable as are devices equipped with multiple sensors or recording pens.

Records are a significant part of the cleaning and sanitizing process. All CIP charts are to be retained by the plant for a minimum of three (3) months. This includes records for cleaning and sanitizing of all plant product processing equipment.

It is also recommended that log records are maintained of manual cleaning operations. This will enable plant quality control staff and the regulatory agency to validate plant cleaning/sanitizing operations.

The Ordinance requires that the regulatory agency review and initial CIP charts during each inspection.

Milk tank trucks shall be tagged or log book records maintained to verify each time the tanker is cleaned and sanitized. The wash tags shall only be removed at the plant or receiving station where the tanker is next cleaned and sanitized. These records shall be retained for 15 days for regulatory review.

Note: Cultured product storage/processing vessels may not have to meet the 24 hour emptying and cleaning requirements if their process demands extended periods of storage, however, a plant should package cultured product within 24 hours of breaking the curd.

Note: Cultured product storage/processing vessels may not have to meet the 24 hour emptying and cleaning requirements if their process demands extended periods of storage, however, a plant should package cultured product within 24 hours of breaking the curd.

Recent changes in the PMO permit the restricted use of condensing water from milk evaporators and water reclaimed from milk or milk products as follows:

1. Pre-rinsing of the product surfaces where pre-rinses will not be used in food products, and
2. Cleaning solution make-up water; provided that for EITHER USE, ITEMS #3-11 OF APPENDIX D, V OF THE PMO 1 ARE MET..AND..
   - o There is no carry-over water from one day to the next OR,
   - o The temperature of all water in the storage and distribution system is maintained at 63o C (145o F) or higher by automatic means, or
   - o The water is treated with a suitable, approved chemical to suppress bacterial growth by automatic methods prior to entering the storage tank, AND
   - o Distribution lines and hose stations are clearly identified as “limited use reclaimed water”, AND
   - o Water handling practices and guidelines are clearly described and prominently displayed at appropriate locations within the plant, AND

- o These water lines are not permanently connected to product vessels, without a break to the atmosphere and sufficient automatic controls, to prevent the inadvertent addition of this water to product streams. NOTE: Recovered water may be used as boiler feedwater for boilers not used for generating culinary steam or in a thick, double walled, enclosed heat exchanger.

3. Appendix D, V requires in Items 3 - 11, among other things a turbidity meter, chemical OD test, monitoring devices too automatically divert, organoleptic quality including weekly samples and records, allows addition of approved bacterial suppressants, and adequately constructed and protected storage vessels.

CLEAN-IN-PLACE (CIP)

The recommended steps in a clean in place (CIP) operation may involve the following:

1. Remove those items that require manual cleaning such as fill tubes, manhole gaskets, plug valves, etc.
2. Provide physical breaks between any circuits or tanks containing product.
3. Pre-rinse or flush thoroughly with cool water not to exceed 80oF.
4. Discard pre-rinse water, flushing until relatively clear.
5. Circulate an effective detergent solution throughout the circuit for the period of time necessary to remove the residues in the circuit.
6. Circulate a rinsing water.
7. Circulate an acid detergent when needed followed by another rinse.
8. Sanitize immediately before use.

Most modern plants have installed clean-in-place systems throughout the plant. Prior to this, plants cleaned all their processing equipment, including tanks, vats, pumps and lines manually which involved complete disassembly, manually brushing with a cleaning solution, rinsing, reassembly and finally sanitizing. This was of course all highly labor intensive.

In the early 1950's, companies such as Tri-Clover, Cherry-Burrell, and Alloy Products Corporation developed special CIP joint/gasket assemblies which were conducive to CIP methods.

In the mid 1960's, clean in place systems led to the use of all-welded product piping systems, automatic (air operated) valving systems, large milk storage tanks or silos designed to be cleaned mechanically, and the installation of properly supported and self draining product lines (recommended minimum of 1/16 inch per foot slope).

The process of circulation cleaning, which was first applied in the dairy industry in the 1940's, involves the continuous pumping of rinse (and often times pre-rinse) and wash solutions throughout the milk product circuits for the purpose of the removal of milk and milk residues rendering the inner surface clean to sight and smell.

Now however, there may be separate CIP circuits for different processes, product lines, or product integrities. (raw vs.. pasteurized) and many plants have installed modern computer controls to monitor their cleaning processes.

The CIP cycle begins at the "make-up" tank containing either a clean water rinse or dairy detergent wash solution. Both chlorinated alkaline and "hot acid" cleaners or caustics are used in the cleaning solution.

The solution is circulated by using relatively higher velocity pumps capable of circulating the solution throughout the circuit, then returning it to the cleaning reservoir.

Systems are monitored for temperature which is required under the Ordinance. Many modern systems also are monitoring , controlling and recording the levels of, pH concentration of cleaning detergent and sometimes flow rates.

Temperatures shall be measured and recorded in the return solution line near the make up tank, however other means of measuring, monitoring, and recording CIP solutions may be used if reviewed by the FDA and found acceptable by the local regulatory agency.

Product storage tanks (except silo tanks) cleaned using circulation from the tank outlet through a pump which returns the solution to the tank spray ball would not have to be equipped with a recorder, providing there are no product lines or other attached CIP circuits involved.

Special attention must be given to the adequate venting of large milk storage tanks including milk silo tanks.

One process engineer has stated that the provision of two, unobstructed three inch openings will adequately vent any tank (horizontal or vertical) using temperatures of<150o F. This is not a hard and fast rule but can be used as a guideline in considering adequate venting of CIP operations for tanks.

Before beginning a CIP of larger tanks and silos, the proper CIP venting door device should be installed in the manhole opening to provide adequate space for air movement resulting from extreme differentials in temperatures.

The basic components of any CIP system will include the following:

1. Permanently installed product piping and air operated valves
2. CIP solution make-up tanks
3. CIP pumps
4. CIP supply and return solution piping
5. Spray devices(either permanently mounted or drop- type)
6. Solution collection manifolds
7. Chemical feed systems and equipment
8. The CIP control/monitoring systems and necessary recorders. CIP temperature recorders must meet all requirements as listed in Appendix H, page 207 of the Ordinance.

WATER CHARACTERISTICS AND AFFECTS ON MILK PLANT CLEANING OPERATIONS

Water hardness is the term applied to water supplies characterized by either a high mineral content, usually high in calcium or magnesium bicarbonates, or magnesium chlorides, and sulfates. Generally water hardness is precipitated by most alkaline materials but not by heat.

Some plants may need to condition their water by either selecting the proper detergents or in more severe hard water cases may need to install ion exchangers to remedy the problem.

Water over 100ppm hardness apply in the latter cases and usually softeners are required for boiler water or cooling tower feed water supplies.

The 3-A Accepted Practices for Permanently Installed Sanitary Product Pipelines and Cleaning Systems require that "All connections between the solutions circuit and the product circuit shall be constructed as to positively prevent the commingling of the product and solution during processing".

Cleaning efficiencies may be enhanced by introducing pressurized air into the system and most systems are fitted with a method to eliminate entrapped air.

Some systems are designed to recover sanitizing solutions for use as a system pre-rinse on the following CIP cycle.

Vertical or silo tanks are cleaned not by sprays but rely on introduction of the cleaning solution at the top of the tank onto a diversion disc or plate which deflects the solution to make contact with the interior tank surfaces.

Also, the tank must be so fabricated and installed to be self draining and to have a fast flushing action across the bottom of the tank.

No matter how sophisticated the CIP system may be, there will always be a need to clean those portions of the equipment not exposed to the cleaning process. Examples of this would be the agitator paddles in silo tanks, air blow fittings, plug type valves, tank venting devices, etc.

Clean Out of Place (COP)

This is the term used for those cleaning operations involving the removal of small sections of piping, valve parts, filler parts, and other small appurtenances that are not normally cleaned in place and placing them into a cleaning vat. The vat is equipped with a heat source, adequate dairy cleaner or caustic, normally steam injection, and a recirculating pump. Recording devices are not required on these units, although many plants have installed them to provide cleaning records as a part of their overall quality program.

Manual Cleaning Operations

Regardless of the automation and engineering built into a plants operation, there is usually a need to manually clean the contact surfaces of the many small parts used in a normal days operation.

This may include filler parts, valves, filler surfaces (mandrels, shields, guides, etc.), small vats, cultured product packaging equipment appurtenances, tanker pumps, fittings, and valves, culture room product handling utensils and equipment (processing vats, curd cutting knives, whey drainers, steam cookers, curd movers and stirrers, etc.).

Manually washed fillers require hand brush washing of the filler bowl, gaskets, the product supply line, ells, and bushings.

Since brushes are used in manual washing operations, they must be of the non-absorbent, nylon or plastic bristled type and designed to not retain soil (not recommended to use hollow body or handled type) and be quick to dry. Brush integrity must be maintained so that brushes used for floor drains or similar surfaces are not used on contact surfaces. This may be accomplished by color coding, marking, etc.

Utensils should be cleaned using a two compartment wash and rinse sink. Sanitizing with chemicals must be accomplished using a third treatment vat, unless heat is used for sanitizing.

The use of absorbent items, such as rags and sponges should be eliminated to reduce the potential of spreading microorganisms throughout the plant environment.

Separate brushes should be used for product and not-product surfaces and the use of wooden handled brushes, tools, paddles, etc. should not be used in production areas.

Particular attention must be given to cleaning underneath gaskets, "o-rings", and other small orifices in which residue and bacteria may accumulate.

Special emphasis should be given to the following when evaluating the effective cleaning of milk contact surfaces:

- AUP air blow devices/fittings;
- Storage tank agitator blades, shafts, fill inlets, vents, gaskets, shaft "O" rings;
- Filler valves, tanks, small parts (springs, screens, gaskets and "O" rings, vents, drain valve and lines, underside of shielding,);
- Plug valves, "butterfly valves";
- Fruit feed pumps, (varigators), fruit pumps;
- Homogenizer stand pipes;
- Reclaimed product surge tanks, milk cans, lines;
- Silo manhole doors (inside surfaces, sampling cocks, tank surface on inside below door opening);
- Silo vent caps (mold, slime);
- Liquefier (powder mixing tanks) outlet valve and connecting piping;
- Raw receiving lines and fittings;
- Product pumps ("O" rings, back plates, covers);
- Filter housings, springs, gaskets in raw receiving area;
- Air eliminators in receiving areas, air "burps" on product and CIP lines;

- Pasteurizer regenerator plates (points of greatest temperature differential);
- vacuum breakers, single stem flow diversion valves;
- Tank fill goose necks; and In-door sample valves.

CHEMICALS USED IN MILK PLANT CLEANING OPERATIONS

Alkaline Detergents

Used to neutralize, break-up, and suspend soil in the cleaning solution. Alkalines have been termed the "guts" of the dairy cleaners and are usually termed generically as "caustics" and or alkaline chlorinated cleaners.. Chemically they are sodium hydroxide (NaOH, OR CAUSTIC SODA), potassium hydroxide (caustic potash), sodium carbonate (soda ash) or sodium hypochloride (NaOCL) and sodium silicates and have a pH higher than 7. They attack the fat and protein residues on all types of dairy processing equipment.

They are generally combined with a water conditioning additive such as liquid phosphate under hard water conditions and usually contain surfactants or wetting agents which enhances the cleaning action.

The water properties must be considered, since most cleaning solutions are made up of 99% water and the amount of neutralizers used should be customized to meet the individual demands. These detergents condition the water to prevent scum formation, and help to prevent saponification (the chemical reaction which converts the actions of alkali and fats into soap and glycerin) during the cleaning cycle.

Most alkaline detergents contain chlorine which also breaks down the fats and proteins.

Acid Cleaners

Acid detergents are those with a pH of lower than 7 and react with mineral deposits which have accumulated on milk contact surfaces. These acid cleaners may contain either nitric, phosphoric or a mixture of both acids; however phosphoric acid based cleaners are the most widely used in the U.S.

Milk and water mineral deposits (termed milk stone) become hardened and layered on the equipment surfaces and provide excellent surfaces on which "biofilms", which are adherent macro-colonies of bacteria, may thrive. Milk stone may consist of milk solids, calcium, magnesium, iron, sulfates, etc.

Acid based detergents (or rinses) do not saponify. They do, however, tend to neutralize any alkalinity and react with the milk minerals which are then carried away in the rinse cycles. Acid rinses will usually provide a bacteriostatic condition, however the equipment must be effectively cleaned with an chlorinated alkaline based detergent prior to using the acid rinse.

Using an acid detergent without first removing the fats and proteins with a chlorinated alkaline detergent will result in fixing the protein soil to the surface.

Acid rinses may also tend to protect against corrosion and when used in proper strengths will not attack rubber parts.

MILKSTONE White to yellow Minerals from milk Acid wash Regular and proper cleaning procedures coupled w/ acidified rinse FAT/ GREASE Hanging water droplets Greasy (white) appearance 1. Low water temperatures 2. Improper detergent concentration 3. Regular use of acids in place of alkaline detergent 1. Proper water temperature 2. Correct concentration of alkaline detergent Regular and proper cleaning procedures coupled w/ acidified rinse MINERAL (Calcium, Magnesium) White (water- stone) chalky to gray 1. Rinse too hot, drop-out of minerals from water supply 2. Failure to use acid detergents 3. No acidified rinse 4. Alkaline detergent used cannot handle hard water at present concentration Acid wash 1. Acid wash 2. Alkaline detergent used has good water conditioning 3. Water softener or treatment

The following is a list of terms commonly used when referring to the properties of detergents used in the food industry.

1. Emulsification - The spontaneous dispersion of substances (proteins, gross soil particles, etc.) when brought in contact with water.
2. Peptize - The mechanical action combined with a surfactant which results in two immiscible liquids to form a stabilized colloid.
3. Saponification - The combination of fatty acids and oils with alkali to produce soap.
4. Sequestering and Chelating - The prevention of precipitation of hard water constituents by keeping them in suspension as stable and solid compounds.
5. Wetting - Lowering of surface tension which permits cleaners to break the bond between the soil and surfaces.

The Ordinance requires that cleaned milk contact surfaces of equipment be sanitized before each use and that aseptic processing equipment shall be sterilized before use and re-sanitized or re-sterilized in all instances whenever contaminated.

It must be stressed that chemical sanitizers are not effective unless all product surfaces are effectively cleaned prior to their application.

For our purposes only conventional milk processing sanitizing will be addressed in this section.

All multi-use containers, equipment, and utensils must be sanitized before use. The method of sanitization may be either by chemicals or the heat method.

Adequate sanitization may be accomplished by using one of the methods as described in the following table.

Chemical Sanitizers: General Information

GENERAL INFORMATION - The Code of Federal Regulations (CFR's), Title 21, Part 178, Section 1010, Page 316-323, lists all the acceptable sanitizers available for use in food plants. This section list the chemistry and general guidelines for use for each sanitizer, including the recommended strengths. Federal codes also prohibit the use of any chemical in a strength higher than that recommended to accomplish its intended purpose in an effort to reduce the possibilities of harmful chemical residues in food products.

1. Chlorine

Chlorine is the most common chemical sanitizing agent used in the milk industry. One of the more common forms of chlorine is sodium hypochlorite which is the soluble salt of hypochlorous acid combined with sodium chloride. This is the stable form of a liquid chlorine solution however has a limited shelf life.

Granular chlorine sanitizers are based upon the salts of an organic carrier which contains releasable chlorine ions. When blended with water, sodium hypochlorite is released and the hypochlorous acid is the active antimicrobial for all chlorine sanitizers.

The killing rate of chlorine is greatly influenced by the pH of the solution. For instance at a solution of 25 ppm chlorine will deactivate organisms at pH 4 in 15 seconds, while a pH level of 10 would take up to 10 minutes. Temperatures of the solution also play an important role in chlorine effectiveness rates. Optimum temperatures of chlorine range from 75°F - 100°F.

Chlorine solutions deactivate the microorganisms by affecting the cell wall transport mechanisms, causing the formation of chloramines with the amino acids of proteins. This denatures the proteins within the enzymes causing intracellular damage and destroys the microbe.

Chlorine sanitizers are generally used in recirculated cleaning systems for cleaning small parts, milk tank trucks and environmental surfaces.

2. Iodine

Iodine sanitizers used in milk plants are usually in the form of iodophors. Iodophors are simply iodine which are combined with non-ionic wetting agents (surfactants). Iodophors, when diluted for use in the proper concentrations, have a low pH value (pH 2.6 - 5.0) which enhances their germicidal qualities. This along with increased temperatures (up to 120°F) makes them a popular sanitizer for use in milk plants.

Iodine interferes with the intracellular protein in a similar manner to that of chlorine, with the sulfhydryl, amino or carbonyl protein groups being affected.

Iodine sanitizers are used extensively in filling/packaging machines and areas, culture processing equipment, drop hoses and hand dipping stations.

3. Quaternary Ammonium Compounds (Cationics)

These sanitizer agents are commonly known as "Quats" and are highly effective in the denaturing of cell (bacteria) proteins which leads to their destruction.

Bacterial cells walls are electrically charged and can possess a net positive or negative charge largely depending on the acidity of the solution in which they are found. Moreover, each type of bacteria have a different cell wall composition. Generally bacteria grow in the optimum pH range of 6.0 to 8.0, and are generally negatively charged at this growth range. The cationic sanitizer, having a positive charge, attracting their opposite charge (negative). This interaction between the opposite charges tends to produce drastic effects upon the cell wall and will cause fluid leakage and eventually cellular rupture. In concentrations of 50 -100 ppm cationic surfactants will not only denature the cells proteins, but also totally inactivate the enzyme and actually alter the cell wall permeability.

Quats are generally used in case washing areas and other environmental surfaces, but are rarely used in CIP and plant equipment sanitization where culture processes may be damaged.

4. Chlorine Dioxide

This chemical sanitizer has become more widely used in the dairy industry, predominately in the sanitizing of environmental surfaces of equipment, floor drains, and other areas to greatly reduce the microbial load in these area. Recommended (CFR) effective uses of this sanitizer for contact surfaces must not exceed 200ppm and a recommended strength of 100-200ppm has been established. The CFR describes the formulation for this sanitizer by either metering a concentrated chlorine dioxide solution into potable water or by acidification of an aqueous alkaline solution of the oxychloro species.

5. Acid Sanitizers

Aqueous solutions containing one or several of the acceptable acid sanitizers has been effective in the deactivation of microbes. The active acid ingredient may be sulfonic, phosphoric, de-cationic, cyanuric or a blend of the above to provide not more than 100ppm of available halogen.

Note: Trichoromelamine based sanitizers are not approved for use on milk contact surfaces.

Acid sanitizers are used primarily for large (silo) milk storage tanks, milk tankers and other large milk storage vessels within the plant.

STORAGE OF CLEANED CONTAINERS AND EQUIPMENT

"After cleaning, all multi-use milk or milk product containers, utensils, and equipment shall be transported and stored to assure complete drainage, and shall be protected from contamination before use".

Once the milk processing equipment has been properly cleaned special care must be taken to prevent contamination prior to the next use. Cleaned equipment and/or utensils must be stored to be protected from splash, dust, insects, overhead dripping, condensation, unnecessary handling, or any other possible source of contamination.

Multi-use utensils, containers and equipment must be stored above the floor on clean racks or other storage devices, inverted for protection and drainage. Care must be taken not to contaminate the clean equipment or utensils by splash from floor, wall or ceiling cleaning operations.

Portable packaging machines must be stored in areas where contamination is less likely, preferable in processing equivalent rooms or acceptable dry storage rooms and should be protected in some manner such as covering with clean plastic sheeting, etc.

Clean equipment must never be stored on dock areas, in machinery areas (boiler or sweet water rooms), or in out-buildings which are not vermin proof or otherwise unsuitable for food processing equipment storage.

Pros and Cons of Chemical Sanitizers

Chlorine Advantages

1. Economical
2. Effective against most bacteria, spores, phages
3. Readily available in liquid or granular form
4. Not affected by hard water salts
5. Test kits available

Chlorine Disadvantages

1. Corrosive
2. Effectiveness is pH dependent
3. Strength dissipates under storage, heat or light
4. Skin irritant, objectionable odor
5. Hard on rubber parts

Iodine Advantages

1. Color indicative of strength
2. Activity less affected by organics
3. Effective against wide variety of non-sporeformers
4. Solution not affected by hard water salts
5. Not a skin irritant
6. Stable, long shelf life
7. Non-corrosive
8. Test kit available

Iodine Disadvantages

1. Effectiveness decreases with increased pH
2. May discolor equipment

3. Less effective against spores
4. Should be used above 120o F.
5. May cause product off flavors

Quaternary Ammonium Compounds (Cationics) Advantages

1. Active against wide range of organisms, especially gram positive slime-formers, thermodurics
2. Low toxicity, odorless, colorless
3. Non-corrosive
4. Stability excellent
5. Heat stable, long lasting
6. Less affected by pH
7. Non-irritant to skin

Quaternary Ammonium Compounds (Cationics) Disadvantages

1. Undesirable in cultured dairy product applications
2. Low activity against gram negative and coliform organisms
3. Loses effectiveness with anionic detergent contamination.

Anionic Acids Advantages

1. Non-corrosive
2. Unaffected by hard water, temperature, or organic contaminants
3. Rapid activity
4. Effective on wide range of microbes(thermodurics, phages, and yeasts)
5. Controls milk-stone

Anionic Acids Disadvantages

1. Increase in pH level decreases effectiveness
2. Cannot be used with alkaline detergents
3. Is less effective against thermo- durics, spores, psychrophils
4. Relatively expensive
5. Leaves foam residual in CIP applications

SUMMARY

The modern dairy industry is very complex unlike old times when it was limited to basic products like milk, butter, cheese and milk powders. But now, dairy manufacturers are providing a sophisticated and ever-changing range of products to customers all over the world. This has put pressure on existing plants to produce more with existing capacities, assure safety of the products, lower production cost and upgrade the products to higher quality standards. Recently, implemented Food Safety Standards Act has also put pressure on the Indian food industry to maintain and improve high quality standards. The changes and pressures in the dairy industry are demanding changes in the technology used in cleaning and sanitation. The dairy industry today needs to have products which can clean and disinfect in the shortest period of time, maintain highest quality, hygiene standards and ensure that the environmental norms are fully complied with. Cleaning and disinfecting are very complex tasks and depend on various factors like (Material of Construction) MOC of equipment, surface properties, soil composition, bonding between soil and surface, processing parameters, limitations in cleaning in terms of maximum temperature and time availability, acceptable residual chemical limit after cleaning, required degree of cleaning and disinfection, etc. These factors demand production units to apply specialised approach to maintain proper cleanliness of various surfaces, especially food contact surfaces. If surfaces in a food production plant are not cleaned to an acceptable level, they may act as a potential source of microbial growth. Further, improper cleaning may increase the utilities consumption viz. steam, water, electricity, etc. The use of specially formulated chemicals for cleaning and disinfecting is an emerging trend in Indian Dairy Industry. Formulated chemicals are customised as per the requirement. For example, chemical with more chelating agent (to help mineral soil removal) is required for high heat-treated surfaces and for cold surfaces, chemicals with more emulsification capacity are required. There may be one or a combination of criteria for selecting a formulated chemical, like faster cleaning with savings in water & energy and cleaning and disinfection in one stage. In dairy industry, hygiene of CIP (cleaning in place) plays a very important role. There are seven steps involved in a normal CIP-hygiene cycle starting from pre-rinse and ending with final disinfection.

6

Water Supply and Waste Water Treatment in Dairy

Large quantities of water are required in any processing dairy on a daily basis for production and other technical purposes. The quantum of water required is calculated mostly on per thousand litres of milk processed and in some cases, the absolute consumption of water per day or per annum is calculated.

The water consumption in a dairy is usually 1000 - 5000 litres per 1000 litres of milk processed. But the actual requirement may vary between different dairies and is mainly dependent on the production process or the production profile, supplementary or auxiliary processes such as generation of steam, design of the refrigeration mechanism involved including domestic use and finally multiple use or reuse.

We all know that water is a precious or very valuable basic resource that belongs to one and all and is not available in unlimited amounts; it is therefore an important cost factor, bearing much significance and hence should be carefully utilized. The design of the water supply system should be such that consumption for each production department and auxiliary processes can be more or less determined in an accurate manner.

This helps to avoid major losses or wastages and paves the way for scrupulous and judicious utilization of water.

WATER SUPPLY

The uninterrupted operation in any dairy depends among other things, the constant availability of the required quantum of water. Hence uninterrupted and potable quality water is a prerequisite in a processing dairy. The possible source of water supply for any dairy include

- Own water supply
- Water supplied through the municipal water system and
- Mixed supply (own water and municipal water; i.e. a combination of the above two)

Regarding the own water supply or self-supply of water, the major resources are well water, ground water and / or surface water which can be used with limitations. But the own supply system requires a great deal of investment for the construction of installations like well, machines, treatment plants and others and it requires a continuously controlled operation.

Most of the time, the water quantities from the own supply or self-supply system are not sufficient to meet all requirements, and hence calls for other alternatives such as such as a mixed water supply system or water from the municipal water source is required.

The municipal water supply is controlled by the local authorities, and they undertake analysis of water samples from the plant.

In the case of self-supply system, water is usually pumped from a well or deep bore well into a large storage tank or basin, which is usually kept covered to keep the water cool and prevent air borne contamination. Generally the capacity of the storage system is equal to 1-2 days of total water consumption.

QUALITY OF WATER USED IN DAIRY PROCESSING

The water received in the dairy shall be primarily fresh and of potable in nature. The drinking water supplied to a dairy in its original or treated state shall be suitable for human consumption without causing any health risks and shall be without any objections in taste, odour and appearance.

Fresh water that has been used in heat exchange circuits for cooling purposes and has absorbed heat can be used as warm water, e.g. for cleaning purposes. Warm water is fresh water coming off from cooling installations. Quality requirements depend on the utilization, whether water is used in production (as cooling water, water for cleaning, boiler feed water) or for social services (bathrooms, kitchens). Water used in a dairy must meet microbiological standards, so that germs do not have any effect on the manufacturing processes or on products. It must be free of inorganic or organic sediments and must be colourless.

Certain requirements regarding temperature and chemical composition must also be met; organoleptic quality must be in line with that of drinking water.

Water For Production Purposes

This includes water that is in direct contact with or incorporated in milk or milk products. Examples are water for the production of long shelf life products, for the manufacture of lactose, butter wash water, water for cheese production and water as a solvent for product preparations. This water must meet drinking water quality standards.

Cooling Water

Cooling water, that is used directly for cooling of milk and milk products must meet drinking water quality standards but can be used later as warm water for cleaning processes or in heat recovery circuits. However when surface water that does not meet drinking water standards is used for cooling purposes, it must be discharged as wastewater.

A special requirement for cooling water is that the hydrogen carbonate hardness should not exceed 6° (German hardness); otherwise sediments and incrustations can form on the walls of the equipment and inside the pipelines. If cooling water is recirculated for heat recovery in open cooling installations, another aspect to be considered is that the original components in the water are concentrated by evaporation and are deposited on the heat exchanger surfaces.

QUALITY STANDARDS FOR DRINKING WATER USED FOR MANUFACTURING PROCESSES IN DAIRIES

Water For Cleaning Purposes

Water used for cleaning of machines, installations, pipelines and storage tanks must also meet drinking water quality standards. If permitted by the authorities, used water can be utilized for cleaning purposes if

(a) The original quality has been drinking water.
(b) It has been used only in closed circuits.
(c) Storage tanks for used water are closed.
(d) Used water from storage tanks has been chlorinated.

The objective is to utilize used water for all cleaning purposes in order to reduce fresh water consumption and to recover energy. For cleaning premises and streets, other types of water can be used.

Boiler Feed Water

There are no hygienic requirements for water that is used in the manufacture of steam in boilers. An exception is the manufacture of "culinary steam". In the UHT processing of milk, the steam produced is called culinary steam and for its production, the water quality standards must be equivalent to the drinking water. Boiler feed water does not have to meet the standards for nitrate and chloride content. However, carbonate and sulfate hardness must be very low.

Hardness in water is generally caused by dissolved calcium, magnesium and iron salts (especially hydrogen carbonates and sulfates). Total hardness of water is composed of temporary or carbonate hardness and permanent or sulfate hardness, which is due to calcium and magnesium sulfate. It is possible to remove carbonate hardness by boiling.

Fresh Water Needs for Dairy Farms

Dairy cows and milking facilities require a reliable high quality water supply. Dairy production is changing with larger production units being planned and constructed. These dairy facilities require a water supply that can deliver high quality water to the cows at all times in sufficient amounts to meet both physiological and sanitation needs of the facility.

The data reported in the literature for water consumption for dairy facilities varies considerably. For example, in three different studies the water needs for lactating cow consumption, and milking center cleaning range between 40 to 50 gallons fresh water per cow per day. In these studies the free stall units were flushed with recycled lagoon water in all areas except the milk parlor and milk room. This concept reduces the amount of water that is introduced into the waste management system.

Texas data illustrates that water use on eleven dairy farms varied widely, depending on the type of manure removal system and other factors. The greatest variable was the use of sprinkler cow wash systems and/or flush systems versus manual manure removal. The total amount of fresh water used for sanitation and manure removal averaged 39.6 gallons per cow per day. The mean value for each farm ranged from 12.3 to 69.2 gallons per cow per day. The water used for cattle drinking water troughs on seven of the dairies studied averaged 28.7±12.0 gallons per cow per day. Thus the Texas data average fresh water use per cow per day for sanitation and drinking was 68.3 gallons. Their data showed wide differences in water use for facilities that flushed or had cow sprinklers versus those that used manual cleanup. The average water use for milking parlors and holding pens without flush or cow sprinklers averaged 19.9±14.0 gallons per cow per day. In contrast, milking parlors and holding pens with sprinkler cow washers used an average of 47.1±17.7 gallons per cow per day.

Calculating Herd Requirements for Fresh Water

It is possible to estimate fresh water use on individual dairies, even with the wide variation reported in the literature. Estimates for total requirements can be developed for drinking water, wash pen sprinklers (if used), and milk house and parlor. Water needs can be estimated by completing Worksheet 1 (requires Adobe Acrobat) with information found in Tables 1-2.

Drinking Water Needs

Drinking water needs can be determined by using the information found in Table 1 and Table 2. Table 1 shows a typical herd makeup assuming a uniform year around calving, 13 month calving interval, a 33% replacement rate (1/2 supplying a calf), 50% female calves and males sold at birth for a 100 lactating cow herd.

Table 1: Typical Dairy Herd Composition

Category	Number	% of herd	Avg. weight (lbs.)
Milking cows	83	83	1400
Dry cows	17	17	1550
Total mature cows	100		
Heifers			
16 - 24 months	42	38	1050
13 - 15 months	13	12	800
9 - 12 months	19	17	600
5 - 8 months	19	17	400
3 - 4 months	9	8	250
0 - 2 months	9	8	150
Total replacements	111		

Water intakes for dairy cattle are influenced by several factors including dry matter intake, milk production, ambient temperature, and sodium intake. However, Table 2 provides a good estimate of water intake based upon size, milk production and temperature. The data from the water intake table when matched with the herd makeup table will estimate herd drinking water.

Table 2: Dairy Cattle Water Intake

Weight (lb)	Milk (lb)	40° & below	60°	80°F	
Lactating cows	1400	202	12.0	14.5	17.9
gal/day	602	22.0	26.1	24.7	
	802	27.0	31.9	38.7	
	1002	32.0	37.7	45.7	
Dry cows	14001	9.7	12.0	16.2	
	16001	10.4	12.8	17.3	
Heifers	12001	8.7	10.8	14.5	
	800	6.3	7.9	10.6	
	400	3.7	4.6	6.1	
	200	2.0	2.5	3.3	

1. Maintenance and pregnancy
2. Maintenance and milk production

Wash Pen Needs

The wash pen can be a large user of fresh water on typical dairies. The amount used is dependent on line pressure, sprinkler orifice, number of sprinklers and minutes per day used. Table 2 shows the gallons per minute per sprinkler with different line pressure in pounds per square inch. The number of sprinklers in the wash pen, the minutes used per day and the nozzle discharge will determine water used in the wash pen.

Application of Electrocoagulation Process for Dairy Wastewater Treatment

Food processing industries cause severe environmental problems because of generation of strong wastewater characterized by high biological oxygen demand (BOD5) and chemical oxygen demand (COD). Among the food industries, the dairy industry is the most polluting in terms of volume of effluent generated as well as in terms of its characteristics too, generating about 0.2-10 L of effluent per liter of processed milk .

The dairy industry is generally considered to be the largest source of food processing wastewater in many countries. Water is used throughout all steps of the dairy industry, including cleaning, sanitization, heating, cooling, and floor washing; naturally, the industry's need for water is huge . In general, wastes from the dairy processing industry contain a high concentration of organic material such as proteins, carbohydrates and lipids, high BOD5 and COD, and high concentrations of suspended solids and suspended oil grease. The BOD5 concentrations for dairy factory wastewaters can vary widely depending on the season or product cycle. For instance, it has been reported that cheese factories have BOD5 ranging from 588 to 5000?mg/L, whereas factories producing cream have BOD5 in the range of 1200-4000?mg/L.

All of these require specialized treatments to prevent or minimize environmental problems. Dairy wastewaters (DWs) are also characterized by wide fluctuations in flow rates, related to discontinuity in the production cycles of different products. Therefore, dairy wastewater treatment is very important from environment and water requirement point of view for dairy industry.

Dairy wastewaters are generally treated usually using biological methods such as activated sludge process, aerated lagoons, aerobic bioreactor, trickling filters, sequencing batch reactor (SBR), upflow anaerobic sludge blanket (UASB) reactor, upflow anaerobic filters, and biocoagulation,. Aerobic biological processes are high energy intensive, whereas anaerobic treatment of dairy wastewater reflects very poor nutrient removal, and effluents treated by anaerobic biological processes need additional treatment. On the other hand, the physical/chemical methods that have been proven to be successful are coagulation/flocculation.

In the recent years, investigations have been focused on the treatment of wastewaters using electrocoagulation (EC) because of the increase in environmental restrictions on effluent wastewater. Electrocoagulation (EC) is an electrochemical method for treating polluted water which has been successfully applied not only for the treatment of soluble or colloidal pollutants, such as slaughterhouse wastewater, vegetable oil refinery, dairy industry wastewater, slaughterhouse wastewater, nitrate-bearing wastewater, wastewaters containing heavy metals, and pesticides and phenolic compounds, but also drinking water for fluoride and humic acid removal.

Electrocoagulation involves the generation of coagulants in situ by dissolving electrically either aluminium or iron ions from aluminum or iron electrodes, respectively. The metal ions generation takes place at the anode, and hydrogen gas is released from the cathode. The hydrogen gas would also help to float the flocculated particles out of the water. The electrodes can be arranged in a monopolar or bipolar mode. The materials can be aluminium or iron in plate form or packed form of scraps such as steel turnings and millings.

Dairy Wastewater

The wastewater used in this work was taken from the local dairy factory in Iran with 25000 (mean value) kg milk per day processing capacity.

Although the quality of the effluent was variable, it was colored milk, and typical characteristics of the raw dairy wastewater are presented in Table 1. Samples were collected in polypropylene bottles, shipped cold and kept at 4°C before use. The length of the storage before starting experiments varied from one day to six weeks.

The effluent has been sampled at different times during this study, and the initial characteristics varied with time. This effluent initially contained high concentrations of soluble and suspended materials.

Setup and Procedure

In each run, wastewater (supernatant) after 12?h settling time was poured into the electrocoagulation cell. All experiments were performed in a bipolar batch reactor with six aluminum electrode connected in parallel. Only the outer electrodes were connected to the power source, and anodic and cathodic reactions occurred on each surface of the inner electrode when the current passed through the electrodes. The internal size of the cell was 15 cm × 15 cm × 25 cm with an effective volume of 2000 cm3.

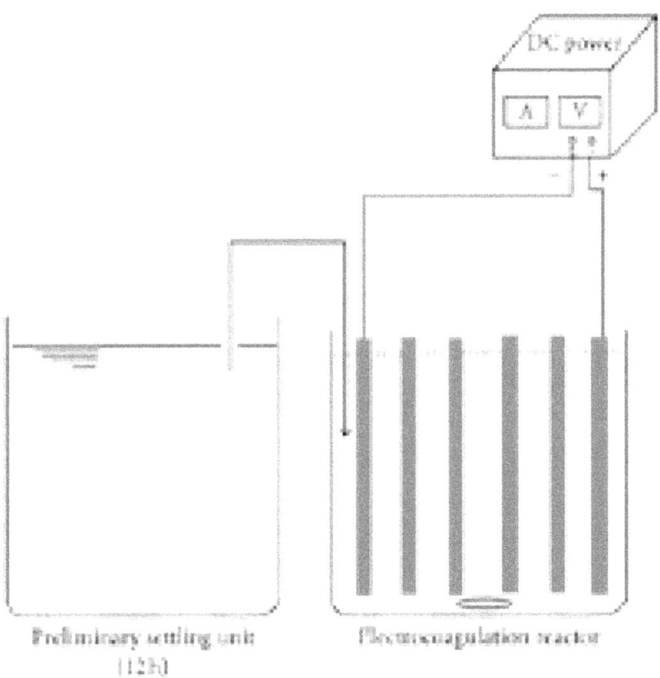

Figure 1: Experimental setup of electrocoagulation unit.

The volume (V) of the solution of each batch was 2?L. The active area of each electrode (plate) was 14 cm × 20 cm (submerged area) with a total area of 280 cm2. The distance between electrodes was 2 cm. A power supply having an input of 220?V and variable output of 0-60?V (10, 20, 30, 40, 50, and 60?V) with maximum current of 5 ampere was used as direct current source. The temperature of each system was maintained at °C. Different samples of 100-150?mL were taken at 15?min intervals for up to 1?h and filtered before being analysed to determine BOD5, COD, TSS, total coliforms, fecal coliforms, and other parameters. During the runs, the reactor unit was stirred at 150?rpm by a magnetic stirrer to allow the chemical precipitate to grow large enough for removal. During electrocoagulation, an oxide film formed at the anode.

In order to overcome electrode passivation at the anode, the electrodes were rinsed in diluted HCl solution (5% v/v) after each experiment and rinsed again with tap water and finally weighted. Also, the electrodes reweighted to calculate sacrificial electrode consumptions. These weights are used in the calculations of the total operating cost. In addition, the electrical energy consumed per unit volume of treated wastewater has been calculated for different experimental conditions. All analyses were conducted in duplicate for reproducibility of the experimental results, and all of the data in the figures and tables were the average ones.

Analytica

COD, BOD5, total suspended solids (TSSs), conductivity, total and fecal coliforms, and pH determinations were determined according to the standard methods (APHA, 2005). COD was measured using COD reactor and direct reading spectrophotometer (DR/5000, HACH, USA). Five-day biological oxygen demand (BOD5) was determined by the manometric method with a respirometer OxiTop system (WTW). The pH and conductivity were adjusted to a desirable value using NaOH or H2SO4 and NaCl and measured using a pH meter model UB-10 (Ultra Basic, US) and a conductivimeter model Cond 3110, respectively.

Results and Discussion

The effects of parameters such as applied voltage and reaction time have been evaluated under specific conditions for a constant temperature.

Wastewater Characterization

Presents the real dairy wastewater characteristics prior to any treatment, after 12 h settling time and the guidelines from Iran for effluent discharge in the sewage urban works. The values of the pollution parameters were lowered after 12?h of preliminary settling time. Also, the comparison of these values showed that the COD, BOD5, and the concentration of TSS, were much greater than those recommended by Iran. Consequently, the dairy effluent needed to be treated before discharge.

Effect of Preliminary Settling Time

Preliminary settling process is a natural treatment method that requires no chemical addition. Although some workers realized the importance of the natural settling process, there is little information available in the literature on the effect of the preliminary settling time on pollutants removal capacity.

In this study, the raw dairy wastewater was allowed to settle in a preliminary settling tank before the electrocoagulation process. The process had an effect on BOD5, COD, TSS, and total and fecal coliforms removals on the first 12 h. COD was reduced from to ?mg/L (resulting in more than 22% COD removal efficiency), whereas BOD5 was reduced in the wastewater from to ?mg/L (resulting in 16%-17% BOD5 removal efficiency). Furthermore, TSS concentration was reduced from to ?mg/L (resulting in more than 57% TSS removal efficiency). Similar results were reported by Amuda and Alade .

Also, data revealed that the effluent of the settling unit is characterized by high load of organic matter. The ratio BOD5/COD of approximately 0.48 indicates that 50% of the COD of this wastewater is easily able to be degraded by biological treatment. Nevertheless, the remainder COD is high, which indicates the necessity of an efficient physicochemical treatment for this wastewater.

Effect of Electrocoagulation Process

Electrocoagulation processes a direct current source between metal electrodes immersed in wastewater. The electrical current causes the dissolution of metal electrodes commonly iron and aluminum into wastewater. The dissolved metal ions, at an appropriate pH, can form wide ranges of coagulated species and metal hydroxides that destabilize and aggregate the suspended particles or precipitate and adsorb dissolved contaminants.

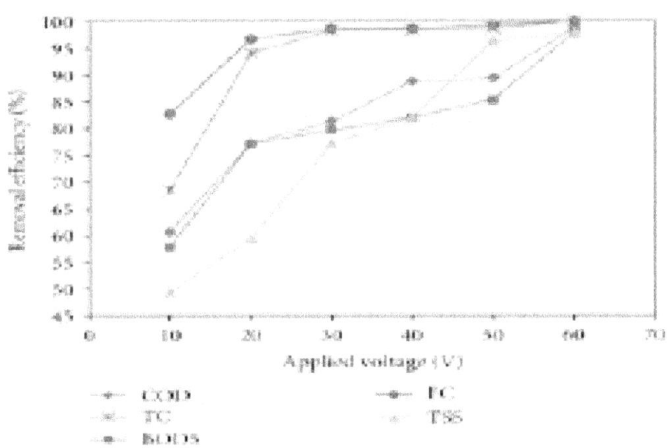

Figure 2: Effect of applied voltage on pollutants removal from dairy wastewater (reaction time = 60 min).

The effect of applied voltage and reaction time on electrocoagulation process of real dairy wastewater treatment were determined. The results of effect of operating parameters on pilot scale electrocoagulation process are shown in Table 2 and Figure 2. The applied voltage is expected to exhibit a strong effect on electrocoagulation process, especially on the BOD5 abatement and COD removal: the higher the voltage, the shorter the treatment. The supply of voltage to the electrocoagulation system determines the amount of Al3+ ion released from the respective electrodes and the quantity of resulting coagulant. Thus, more Al3+ ions get dissolved into the solution, and the formation rate of Al(OH)3 is increased. Also, it is well known that electrical potential not only determines the coagulant dosage rate but also the bubble production rate and size and the flocs growth which can influence the treatment efficiency of the electrocoagulation process.

Illustrate the effect of applied voltage on pollutants removal efficiency at pH of 7.24. Removal efficiencies of more than 97% were achieved after 60 min electrocoagulation for all voltages (10-60 V). Also, it is clear that by increasing electrical potential from 10 to 60?V, the pollutants concentration in the effluent decreased significantly. As an example, COD concentration has decreased from 6114.25 to 2405.96?mg/L (approximately 60.6% COD removal efficiency) after electrocoagulation process with electrical potential of 10?V. Again, by increasing electrical potential to 60?V, the COD concentration in the effluent decreased to 70.9 mg/L in 60?min (approximately 98.8% COD removal efficiency). Also, BOD5 concentration has reduced from 2919.3 to 892.2?mg/L (nearly 57.9% BOD5removal efficiency) after electrocoagulation process with electrical potential of 10?V, and by increasing electrical potential to 60?V, the BOD5 concentration in the effluent declined to 43.45?mg/L in 60?min (around 98% BOD5 removal efficiency). The same trend was observed for TSS concentration in wastewater after electrocoagulation process, so the removal efficiency increased from 49.29% at electrical potential of 10 V to 97.75% at electrical potential to 60?V. Furthermore, as can be seen from Figure 2 and Table 2, the removal efficiency of bacterial indicators (TC and FC) is very high, and efficiency was raised with increase in applied voltage from 10 to 60?V. Maximum removal efficiency (>99.9%) was obtained in applied voltage 60 V, and thus the effluent quality was reached to permissive levels (lower than 1000 and 400 for TC and FC, resp.), and hence from the viewpoint of microbial aspects, discharge of this effluent to environment is safe. Also, minimum removal efficiency occurred in the lowest electrical potential (10 V). This is ascribed to the fact that at high voltage, the amount of oxidized aluminum increased, resulting in a greater amount of precipitate for the removal of pollutants. In addition, it was demonstrated that bubbles density increases and their size decreases with increasing current density , resulting in a greater upwards flux and a faster removal of pollutants and sludge flotation. In fact, the most important mechanisms for removal of pollutants in this process are rapid adsorption of soluble organic compounds and trapping of colloidal particles in "sweep flocs" (Al(OH)3). The results of this study are in agreement with the results of researches done by Nouri et al. on zinc and copper removal from aqueous solutions by aluminum electrodes, Bazrafshan et al. on the efficiency of EC with iron and aluminum electrodes in removal of chromium, Mansoorian et al. on nickel removal from aqueous solutions by EC process with iron-rod electrodes, Njiki et al. on mercury removal from water by EC process with aluminum and iron electrodes, Bazrafshan et al. on humic acid removal from underground waters by

EC process using iron electrodes, Bazrafshan et al. on the efficiency of EC with aluminum and iron electrodes in removal of phenol, and also Bazrafshan et al. on the efficiency of EC with aluminum and iron electrodes in removal of fluoride from aqueous solutions. Nevertheless, Bayar et al. reported that increase in the current density does not cause an expected removal efficiency increase; on the contrary, it can cause a relatively negative effect on it. Also, a similar trend was seen in the study of Holt et al.

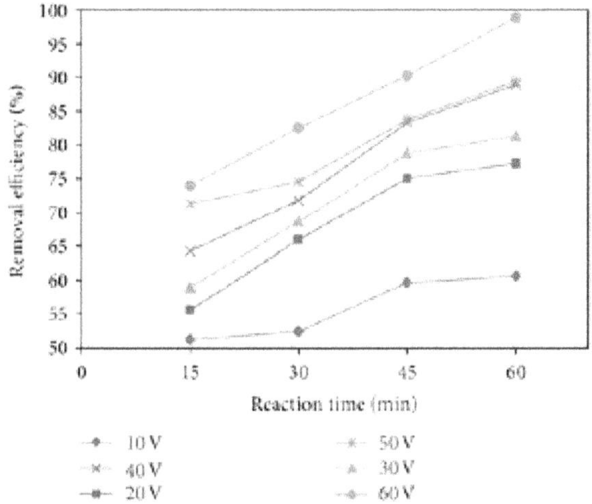

Figure 3: Effect of reaction time on COD removal from dairy wastewater at various applied voltages.

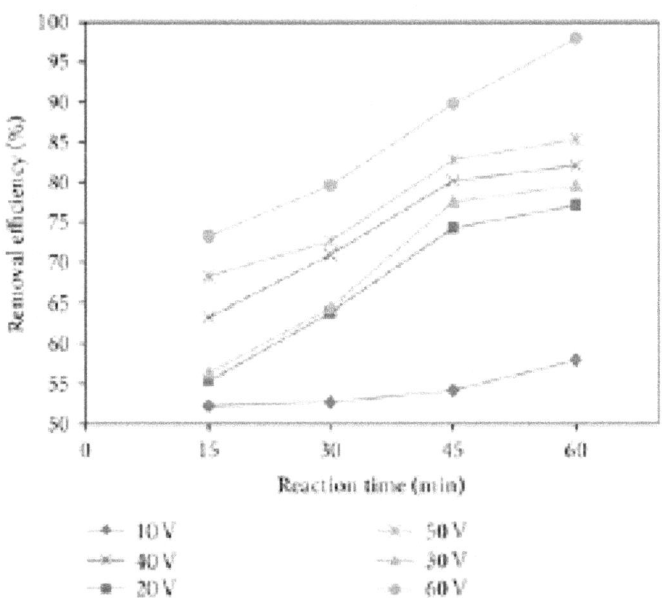

Figure 4: Effect of reaction time on BOD5 removal from dairy wastewater at various applied voltages.

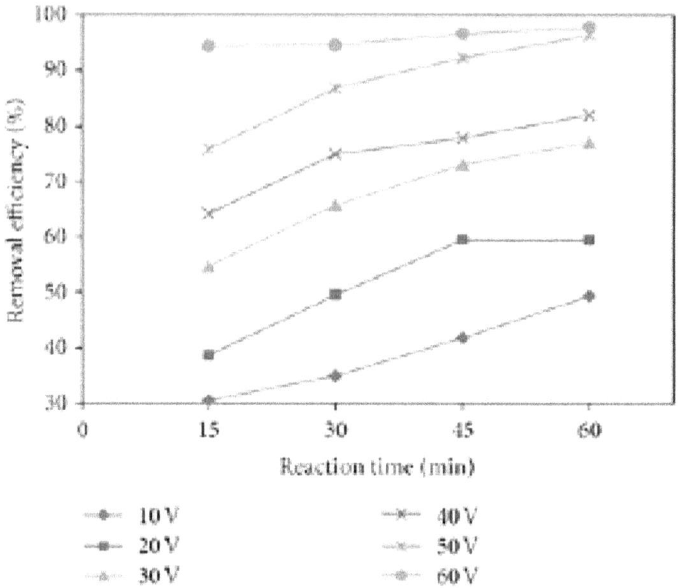

Figure 5: Effect of reaction time on TSS removal from dairy wastewater at various applied voltages.

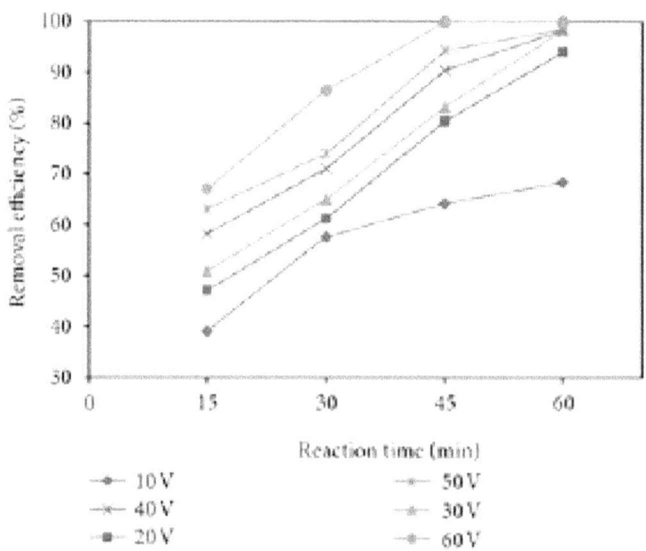

Figure 6: Effect of reaction time on TC removal from dairy wastewater at various applied voltages.

Represent the effect of reaction time on pollutants removal from dairy wastewater at various applied voltages. Theoretically, based on Faraday's law, duration of electrolysis affects the amount of released electrode in a system with aluminum electrodes and determines the amount of produced Al3+ from this electrodes . EC process includes two steps, destabilization and accumulation. The first step is usually short, and the second one is relatively long. With increase in reaction time, both energy and electrode consumption increase and this shows that reaction time is a very important parameter due to affecting the cost effectiveness of EC process in polluted waters.

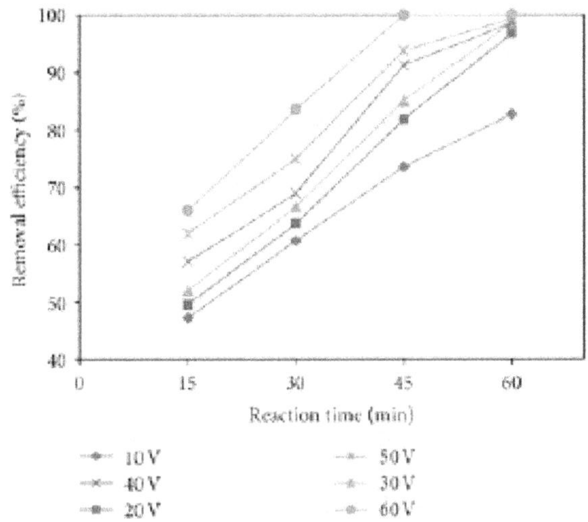

Figure 7: Effect of reaction time on FC removal from dairy wastewater at various applied voltages.

In the present study, as it can be seen from Figures 3-7, more than 50% of wastewater quality parameters were removed in the first minutes (only 15 min), especially for applied voltages 30, 40, 50, and 60 V, while in later minutes, the percentage of removal was low. Also, more than 50% removal efficiency of COD and BOD5 was obtained at all applied voltages at 15 min. Furthermore, as the time of electrolysis was increased, comparable increases in the pollutants removal rate were observed for all applied voltages. On the other hand, for a given time, the removal efficiency increases significantly with an increase in applied voltages. As an example, after 60 min of electrolysis it can be seen that 60.65%, 77.25%, 81.29%, 88.85%, 89.37%, and 98.84% of COD were removed for applied voltages of 10, 20, 30, 40, 50, and 60 V, respectively. Also, as the applied voltage was increased, the required time for the electrocoagulation process decreased. This was ascribed to the fact that at high current, the amount of oxidized metal increased, resulting in a greater amount of precipitate for the removal of pollutants. Similar trends were observed for other quality parameters of dairy wastewater.

The results of this study is in accordance with the results of researches done by Wan et al. on removal of arsenic from water by using EC, Ugurlu et al. on the removal of lignin and phenol from paper mill effluents by EC, and also Merzouk et al. on removal of turbidity and separation of heavy metals using electrocoagulation-electroflotation technique.

Electrical Energy and Electrode Consumption

Electrical energy consumption is a very important economical parameter in the electrocoagulation process. Therefore, for the same operating conditions, after 60?min of electrocoagulation, consumption of energy and aluminum electrode were determined, that is, represented in Table 3. It can be understood from Table 3 that electrical energy and electrode consumption were found to increase with increasing the applied voltage as would be expected in any other electrolytic process. An increase in applied voltage from 10 to 60 V causes an increase in energy consumption from 0.012 to 0.095 kWh/L. Similar trends were seen in the study of Bayar et al. on Poultry slaughterhouse wastewater treatment and Bazrafshan et al. on carwash wastewater treatment by electrocoagulation process .

Also, as it is presented in this table, electrode consumption varied between 0.249 and 2.672?g/L. It can be concluded that the higher the voltage of the system applied, the higher the weight of the electrode consumed in the process. On the other hand, an increase in the applied voltage causes a proportional increase of the consumption of electrode by the system. As the table represents, the electrode consumed at 60?V in the process is much more than that of the process conducted with 10?V. This result is in agreement with the findings of Bazrafshan et al. on slaughterhouse wastewater treatment by electrocoagulation process .

Batch electrocoagulation studies were performed to evaluate the influence of various experimental parameters such as applied voltage and electrolysis time on the removal of pollutants from dairy wastewater. The results of this study have shown the applicability of electrocoagulation in the treatment of real dairy industry wastewater. The treatment rate was shown to increase upon increasing the applied voltage and reaction time. However, increasing the applied voltage caused the energy consumption to increase. Indeed, the highest voltage produced the quickest treatment with an effective reduction of COD, BOD5, and bacterial indicators (TC and FC) concentration. Consequently, it can be inferred that electrocoagulation is a comparatively suitable process for removal of COD, BOD5, and other pollutants using aluminum electrodes to effectively treat dairy industry wastewater.

Treatment of dairy industry wastewater

In a background of natural water resource availability and cost increase, wastewater treatment for water reuse can lower the overall water consumption and the global effluent volume of industrial plants. Like other industries (textile, pulp and paper, tanneries, etc.), the food industry looks at membrane processes for wastewater treatment to produce purified water for recycle or reuse. In the beverage industry, nanofiltration and low-pressure reverse osmosis were used for treating water from bottle washing water to produce water of drinking quality; spent process water from fruit juice was processed using a membrane bioreactor and two-stage nanofiltration for water reuse. Ultrafiltration and nanofiltration were used for water recycling from wastewater of the fishmeal industry. Microand ultrafiltration of surimi process water enabled the recycling of the treated water. A combined nanofiltration-reverse osmosis system was developed to treat soybean soaking water for water recovery. Poultry slaughterhouse chiller tank effluent was recovered using ultrafiltration and nanofiltration for purified water recycling. Cooling water from sausage products was treatedby nanofiltration to produce drinking quality water.

Among the food industries, the dairy industry is the most polluting in volume (generating from0.2 to 10 L of effluent per litre of processed milk) in regards to its large water consumption. Several works focused on the treatment of dairy effluents demonstrated that membrane operation were convenient: microfiltration , ultrafiltration, UF , nanofiltration, NF , reverse osmosis, RO or two-stage operations such as NF+NF or RO+RO . Chmiel et al. showed that NF and RO of low contaminated vapour condensate from milk processing can produce reusable water.

The process waters produced in the starting, equilibrating, stopping and rinsing any of the processing units generate a large volume of effluents (flushing water, first rinse water, etc.), contributing significantly to the total wastewater production. Membrane treatment of these selected dairy wastewaters with the aim of water reuse could simultaneously lower the total water consumption and the effluent production of the dairy plant. The purified water produced by membrane treatment could be reused in the dairy factory as heating or cooling water, as boiler make-up water or for cleaning purposes. Its quality must be, at least, similar to those of vapour condensates, collected during the product drying and evaporation processes and often reused for such applications. If the quality reaches the drinking water standards, the field of applications could be broadened with applications such as cooling water for pump seal or plate heat exchangers where unexpected contact with dairy products may occur, but therefore derogation byFrench authorities will be required.

The objective of this study was to use ROunits for the treatment of several selected astewaters from dairy plants for water reuse purposes.This paper summarises first results of investigations (inquiries, visits and samples collection and analyses) made in 11 dairy plants in order to collect qualitative and quantitative informations about dairy process waters, vapour condensates and practices of water management. Stability of wastewater during storage before filtration was also studied. In previous feasibility studies with a synthetic solution (diluted skim milk), performances of eight NF and RO membranes were compared by dead-end filtration . Two selected membranes (one NF and one RO) were assessed for crossflow experiments: the results agreed with those of dead-end filtration .More complex model solutions, representative of the composition variation of the dairy process waters, were processed by the selected NF and RO membranes .

It can be concluded that, in the range of the study, heat treatment, fat content and whey/milk ratio of the model solutions do not have significant effect on the performances (purified water quality, permeate flux). The study also showed that RO operation gave better water quality than NF, as NF did not provide a better permeate flux. Therefore RO membrane was chosen for the treatment of dairy wastewaters presented in this study. Performances in terms of permeate flux, purified water quality and recovery, milk component removal efficiency were discussed. Scalingup of a RO plant of 100 m3 /d was outlined.

EXPERIMENTAL

Feed solution: dairy wastewater (DWW)

Eleven French dairy plants were involved in inquiries about their wastewaters. These inquiries were completed by visits of eight plants in order to collect qualitative and quantitative information. This study showed that dairy process waters (mainly flushing waters, first rinse waters, etc.) were mixtures of milk products diluted with water without chemicals from the cleaning in place steps. Composition and concentration depend on the dairy plant main production and on the processing units. These selected wastewaters are mainly mixtures of milk, whey and cream with dry matter (DM) ranging from 0.4 g.L!1 to 71 g.L1, fat content (0 to 22%) and heat treatment (no heat treatment to high heat treatment: 130EC, 20 mn).

Eight samples were collected in flushing waters from concentrator or pasteuriser processing of skim milk or whey (from five plants). Their compositions are given in Table 1. DM of the eight dairy wastewaters (DWW) ranged from 0.4 to 54.3 g.L1; this large variation depends on the sampling time (beginning or end of the flushing step). It can be observed that the other characteristics depend on dilution, so that the ratio of concentration to dry matter is in the same proportion than in original fluid. For the flushing step, either drinking water or vapour condensate (almost demineralised) was used. When conductivity and ionic concentration (especially nitrate or chloride) were high related to dry matter, as common sense drinking water was used as flushing/rinsing water: this was the case for DWW B, 1 and 3.

Membrane

A commercially available RO spiral-wound membrane was used in this study, a TFC HR SW 2540 (provided by KOCH Membrane Systems) with a NaCl rejection of 99.5%. It is a thin film composite (TFC) membrane with polyamide active layer and an effective filtration area of 2.5 m2.

Deionised water membrane permeability (at 25EC) was 2.5 L.h!1 .m!2 .bar!1 The membrane pure water permeability was measured with deionised water at 25EC before each experiment. After each experiment, the membrane was cleaned up with a 0.04% w/v of alkaline detergent at pH 11 (Ultrasil 10, HenkelEcolab) for 1 h at 43EC and 10 bar. After cleaning, the membrane permeability to water was checked again. If permeability was not achieved, an additional acid cleaning with HNO3 at pH 2 was made for 1 h at 30EC and 10 bar followed by a second alkaline cleaning step. The membrane was stored in a 0.5% w/v sodium metabisulfite solution.

Operating conditions

The crossflow RO filtration device was designed by TIA (Bollène, France). Transmembrane pressure (TMP) and crossflow flowrate, QR, were set at 20 bar and 800 L.h!1, respectively. Temperature (T) of the solutions was controlled at 25EC. The stainless steel feed tank capacity of the filtration device was 50 L. The volume of wastewater ranged between 130 and 200 L and was well mixed in a storage tank. At the beginning of the filtration, the feed tank was entirely filled with 50 L of solution. The filtration was performed according to a continuous concentration mode with step by step feed addition: 5 L of feed solution (wastewater from the storage tank) was added in the feed tank each time 5 L of permeate were collected (fed-batchmode). The concentrate stream was recycled back to the feed tank. So the volume in the feed tank ranged between 45 and 50 L during the first part of the filtration (as long as feed solution was added). The feed solution was continuously concentrated up to a volume reduction ratio VRR of 7 to 17 depending on the feed volume.

Analysis

Feed, retentate and permeate samples were assessed for physicochemical and chemical analysis: dry matter, DM (accuracy ± 2%); suspended matter (French standard method EN 872, accuracy ±2%); dissolved organic carbon (DOC) (French standard method EN 1484, accuracy ±2%); total Kjeldahl nitrogen (TKN) (French standard method EN 25663, accuracy ±1%); total organic carbon (TOC) (OI Analytical TOC analyser, accuracy ±2%); pH (accuracy ±0.05). COD was determined with rapid test tubes (oxidation with potassium dichromate/sulphuric acid/silver sulphate at 148EC, accuracy ±3%) and photometric measurement (Nanocolor 300D) provided by Macherey Nagel. Cation (Na+, K+, Ca2+, Mg2+) concentrations were measured by atomic absorption (Spectra 300, Varian Associates) and anions (phosphate, NO3!, Cl!) concentrations by ion chromatography (Dionex DX 500). Accuracy was about 1% for all ions. Lactose concentration was determined spectrophotometrically at 488 nm by the phenol-sulphuric acid method with an accuracy of 3%. Lactate concentration was determined spectrophotometrically at 425 nm using the method described by Ling with an accuracy of 5%. Ethanol, acetone and acetoin were determined by headspace gas chromatography according to the method described by Richelieu et al. The conductivity ? was measured (CDM 210, Radiometer Analytical) with an accuracy of 2%.

Water management practices in dairy plants

Investigations were conducted in 11 French dairy plants in order to collect information about consumption, origin and uses of water (in 2001). The water consumption depended on the volume of processed milk, on the water requirements of the different processes and on the practices of water management at the factory. The daily consumption ranged between 800 and 3400 m3 /day with an average value of 1700 m3 /day. When river water or groundwater was used as cooling water in open loop, the consumption could reach 12,000 m3 /day. The volume of processed milk ranged from 330 to 1000 m3 /day, consequently the water consumption varied from 1.2 to 3.4 L per L of processed milk. A low ratio showed a good optimisation of the water management.

The water used at the factories came mainly from three origins: drinking water supplied by water provider, water from the natural environment (river water or groundwater) often treated to be potable, vapour condensates coming from the product concentration and drying processes (also called “cow water”). The origin of consumed water depended on the availability of water, on the type of use (with or without food contact) and on the practices of management in each dairy plant. Fig. 1b illustrates that drinking water from water provider was used in 10 plants and represented from 11 to 75% of the consumption. Groundwater was pumped and used in seven plants after treatment in the dairy to make it potable. However, since 2001, the use of groundater still increases because of drinking water cost increase. Fig. 1b shows that in situ available vapour condensate was used in all except one factories (20 to 48% of the water consumption). This statement is consistent with Daufin et al. who showed that condensated vapours were used in almost all the dairy plants and that the recycled volume can reach 103 to 4×103 m3/day in some cases.

The available volume depended on the size of the concentration and drying unit of the dairy. Inquiries did not provide information on the quality of vapour condensate, so five samples were collected on site and analysed. Physicochemical analyses of the five vapour condensates from skim milk, whole milk or whey drying units are shown in Table 2. Condensates from (whole or skim) milk were low mineralised (?<18 µS.cm!1) and did not contain proteins (TKN <2.1 mg.L!1). In two samples, TOC was even lower than 2 mg.L!1 (threshold value for drinking water in France). Vapour condensate coming from whey was more mineralised (? .100 µS.cm!1) and had higher TOC value (11 mg.L!1).This observation is consistent with data of Stroem and Holmström who showed that generally vapour condensate organic matter content of milk drying processes was lower than that of whey. The objective was to identify organic compounds involved in TOC. Stroem analysed the vapour condensates by gas chromatography in order to identify the organic molecules involved in COD and detected small molecules like ethanol, acetone, acetoine, diacetyl and dimethylsulfide. The five collected samples were assessed for such analyses completed with lactose and lactate determination. In vapour condensates from milk drying processes, acetone acetoine and ethanol were detected and responsible for 20 to 70% of TOC. In vapour condensates from whey drying processes, lactose and ethanol represented 50% and 10% of TOC, respectively. The comprehensive identification of all organic molecules involved in TOC would be difficult to perform because more than 100 small volatile organic molecules were detected in fresh milk without taking into account the new molecules produced during dairy products heating and drying steps .

According to French regulations, only drinking water can be used in direct contact with the food products (for instance: starting of processing units, flushing water, last rinsing of pipes, flat cheese rinsing). Drinking water is also required for applications where unexpected contact with dairy products may occur when there is a risk of leakage (pump seal or plate heat exchangers). So, for recycled water such as vapour condensate, uses are restricted to applications without contact with the dairy products.

Our investigation (2001-2002) demonstrated that applications were:

- boiler make-up water (30-275m3/day): vapourcondensate with conductivity <20 µS.cm!1 andwith often additional treatment (activated carbon or RO) was used in seven dairy plants (up to 160m3/day);
- cooling water in closed loop (70-370m3/day):vapour condensate with conductivity <1500µS. cm!1 was used in five dairy plants;
- cleaning in place solutions (30-700 m3 /day): in about half of the plants, vapour condensate was used for preparation of dilute alkaline or acid solutions an for intermediate rinsing;
- outside washing of equipments (8-250 m3/day): in five plants, vapour condensate permitted to wash floors and the outside of tanksand trucks.

One should note that practices in half of the factories were to use drinking water for such purposes where drinking water quality isn't required. In order to keep such a safe preventive behaviour, reuse of recycled dairy industry wastewater might be proposed as new alternative practice.

Dairy wastewater storage

It was shown in a previous study that during storage of diluted dairy products, natural acidification occurred and that quality of permeate after membrane treatment was strongly affected. Therefore, in order to study the stability of dairy wastewater, pH variation of three samples was followed at 25EC during about 5 days. Due to the sampling and travel time, the delay between collection time (t = 0) and the beginning of storage at 25EC was different for each sample. The initial pH value was quite the same (around 6.5) for the three samples, but after 20 h of storage at 25EC the pH value for whey wastewater decreased by more than one unit whereas pH of skim milk wastewater decreased only by 0.1 unit. After 24 h at 25EC, pH of skim milk wastewater began to decrease strongly. Therefore the stability of milk wastewater was better than that of whey wastewater. These results were checked using milk and whey: natural acidification of whey at 25EC occurred before that of milk probably due to a buffer behaviour of caseins . For milk and whey solutions kept at 4EC, pH remained higher than 6 after 90 h. In addition, during the acidification process, low molecular weight organic molecules (such as lactate and ethanol) were produced by lactose degradation . These by-products lower membrane performance because their rejection was low, inducing an increase of permeate TOC. In conclusion, dairy wastewater needs to be stored at low temperature in order to limit natural acidification, otherwise it shall be treated within a few hours after collection, in order to prevent degradation of milk components.

DAIRY WASTEWATER CROSSFLOW RO FILTRATION

Three dairy wastewaters were sampled and filtered in the crossflow filtration device with the procedure previously mentioned. The volume of feed solution was 130 L for DWW 1 and 200 Lfor DWW 2 and DWW3. Permeate fluxes of RO filtration were plotted vs volume reduction ratio. The flux declined continuously for each experiment during retentate concentration. DWW 1 produced in whey concentrator showed a higher flux than DWW 2 and 3 from skim milk processes, below VRR 5, despite a higher DM content. This lower flux with skim milk wastewater can be explained by membrane fouling due to deposit and/or gel formation of caseins; this gel was previously observed on membrane surface of dead-end filtration experiments with diluted skim milk. For VRR higher than 5, the flux fell very low for DWW 1, perhaps due to a deposit of suspended matter which was very high for this sample and possible precipitation of calcium phosphate (scaling). At the start of the run, permeate flux was higher for DWW 2 than for DWW 3 in accord with DM content. But shortly after beginning, the fluxes were similar for the two wastewaters, despite different retentate concentrations, showing similar fouling resistances. It must be outlined that flux decline and fouling analysis were not obvious because the feed solutions were complex mixtures of milk products which are responsible of both osmotic pressure (salts and lactose) and fouling resistance such as suspended matter deposit with in some cases gel formation (probably due to milk caseins) and scaling (precipitation of calcium phosphate). Classical models (as osmotic pressure model, resistance model, gel polarisation model and combined models) were tested, nevertheless due to the complexity of the feed solution it was very difficult to find a model which describes or predicts quantitatively or even qualitatively the total effect of the solutes on flux decline. This mechanistic aspect should be considered in the future with model solutions.

Performances of the filtration i.e. average permeate flux and TOC of collected RO permeate were given according to water recovery which is a key parameter of the wastewater treatment. Average permeate flux could be determined by integration of instantaneous permeate flux or ratio of collected permeate volume to filtration time. For the three samples, average permeate flux decreased vs. water recovery: average flux was in the range 11-13 L.h!1 .m!2 for about 90% recovery and 10-11 L.h!1 .m!2 for 95%.

TOC of overall permeate increased slowly until 80% recovery, but at higher water recovery, permeate TOC increased significantly up to 2, 3.5 and 7 mg.L!1 for DWW 2, 3 and 1, respectively. As expected, permeate TOC was dependent on the wastewater concentration, about 3 times higher for DWW 1 (DM 9.8 g.L!1) than for DWW 2 (DM 3.5 g.L!1). Inspection of data illustrated that RO operation can reach a water recovery of 90 to 95 % with an average permeate flux around 11 L.h!1.m!2 and with TOC of purified water lower than 7 mg.L!1

Removal efficiency of the RO operation was calculated for TOC, Conductivity, TKN, ions, and lactose. As expected, removal efficiency for RO membrane was high for all the compounds: organic matter removal was very high (>99.8 % for TOC and > 99.5% for lactose), nitrogenous matter removal was around 96% and conductivity removal was about 97% (higher than 95% for multivalent ions and 87% for monovalent ions). Results do not show significant difference between DWW 1 from whey and DWW 2 and 3 from milk. At the beginning of the filtration (at VRR 1), TOC rejection was 99.9% and conductivity rejection around 99.5%. These data were fairly consistent with those obtained previously for a RO treatment of a dairy model process water .

Purified water quality

RO permeate characteristics are given in Table 4 for each wastewater. In all the cases, permeate was low mineralised with a conductivity value lower than 50 μS/cm, corresponding mainly to monovalent ions (Na+ , K+ and Cl!). Organic matter content was even very low, TOC value ranging from 2 to 6.7 mg.L!1. Lactose concentration ranged from 5 to 12 mg.L!1 while lactate concentration was lower than the threshold value of the analysis method. Lactose contribution to permeate TOC ranged from 76% to 100%. Other small organic molecules (like lactate, ethanol, and acetone) could probably contribute to the remaining organic matter content involved in TOC but their identification would be difficult because more than 100 small volatile organic molecules were identified in milk products and their concentrations were too low to be detected by analysis methods. RO permeate characteristics do not show significant difference between permeate from milk products (DWW 2 and 3) and permeate from whey (DWW 1). But results show that permeate TOC is strongly related to wastewater concentration (i.e. dry matter): the higher the wastewater concentration the higher the permeate TOC.

For the three wastewaters, purified water (i.e. RO permeate) quality was compared to vapour condensates sampled in dairy plants. Purified water showed TOC concentration and conductivity similar to those of vapour condensates. For DWW 2, TOC reaches even the drinking water standard. So the purified water produced by RO from selected wastewater could be reused in the dairy plants for the same purposes than vapour condensates (i.e. heating, cooling and cleaning applications).

A RO polishing operation was performed after the first RO stage for the improvement of water quality. Permeate flux was quite constant during the filtration around 50 L.h!1.m!2. TOC rejection ranged between 77 and 81% at VRR 1 only, probably due to the low rejection of the small organic molecules. Illustrates that RO+ RO purified water was quite significantly demineralised (conductivity = 3 μS.cm!1) and that TOC complied with the French drinking water standard (TOC <2 mg.L!1) in two cases and was very close to this standard in the last case (2.1

mg.L!1). With derogation from the French authorities, such purified water could be used for new applications where unexpected contact with milk products can occur.

Scale-up

Scale-up was carried out for a RO treatment of 100 m3 .day!1 wastewater in fed-batch mode. The membrane area was determined under the following assumptions: filtration at 20 bar, 25°C, 800 L.h!1 during 16 h.day!1 with an average permeate flux of 11 L.h!1.m!2 (corresponding to the average permeate flux obtained with wastewater). In order to achieve a water recovery of 95%, the filtration plant will require 540 m2 RO membrane arranged for example with 18 spiralwound 8" elements (30 m2 per element, 3 elements per pressure vessel, 6 vessels in parallel).

Water management practices in 11 French dairy plants show that water was mainly coming from three origins: drinking water from water provider, riverwater or groundwater (often treated to be potable) and vapour condensates. One should note that half of the factories used drinking water for applications where such quality was not required. Dairy wastewater stored at 25EC shall be processed within a few hours after sampling in order to prevent increase of permeate TOC due to degradation.

RO treatment of the selected wastewaters was carried out until 90-95% water recovery was achieved with an average permeate flux around 11 L.h!1 .m!2 . TOC of purified water was lower than 7 mg.L!1 and came mainly from lactose (76-100%). Purified water was low mineralised: conductivity <50 μS.cm!1 corresponding essentially to monovalent ions (Na+ , K+ and Cl!). Quality of purified water was similar to vapour condensates, so it can be reused for the same applications as heating, cleaning and cooling.

Scale-up was proposed: a plant of 540 m2 RO membrane area permits the recovery of 95% of 100 m3 per day wastewater. A polishing step with RO stage could be added in order to get water fulfilling requirements of drinking water.

SUMMARY

Milk-processing plants generate significant quantities of wastewater with relatively high organic matter concentrations on a daily basis. In addition to environmental damage that can result from the discharge of these wastewaters into the natural waterways, the presence of products such as milk solids into wastewater streams represents a loss of valuable product for the plants. This paper presents a review of wastewater management practices employed by six milk-processing plants in Victoria, Australia. In all six plants investigated, milk powder represents a major product. During the milk powder production, water is evaporated, condensed and can be reused for various purposes with a significant impact on water usage. Other major products are anhydrous milk fat, cheese, butter, and UHT milk. The effectiveness of the practices was assessed through two main criteria: first through the water to milk intake ratio, and the waste volume coefficient. Both parameters characterise the plant efficiency in regard of water consumption and water reuse, Information on cleaning chemical usage and recovery was also assessed as part of the review. Significant discrepancies emerge between the plants first due to the products manufacturad and water reuse possibilities available in each plant. Second the type of treatment technologies used for condensate and cleaning solution influences the figures. One of the investigated plants is almost self-sufficient for water, emphasising the benefits gained from the use of technologies like membrane separations for condensate and cleaning solution treatment. In some cases, less cost-intensive technologies such as a clarifier are successful to improve cleaning agent recovery.

7

Energy Supply in a Dairy

The cumulative energy demand for feed supply as calculated for the fundamental ration ingredients.

Table 1 Rearing period per cow in a year in dependence on the replacement rate (growing up period of the heifers is 25 months)

Replacement rate [%]	10	15	20	25	30	35	40	45	50
Rearing period per cow and year [months]	2.5	3.8	5.0	6.3	7.5	8.8	10.0	11.3	12.5

The production of concentrate feeds needs a higher cumulative energy demand per kg dry matter than the production of roughage feed. The proportions between the feeds are different when the cumulative energy demand is related to net energy lactation (NEL) or usable crude protein (CP). Maize silage has the lowest cumulative energy demand for all three reference bases except pasture. Pasture requires in all cases the lowest cumulative energy demand because almost no machines are used.

Besides the production of single feeds, the composition of the diet is important for the energy intensity of dairy farming (Kraatz et al., 2006). With increasing milk yields, the energy requirement of cows increases and it must be met by an increase in the energy concentration of the diet.

The energy demand for feed/kg of milk produced (energy intensity) decreases with milk yields between 4 000 kg and 8 000 kg per cow/year (Figure 1).

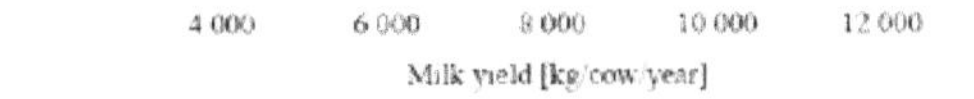

Part of concentrate in the diet (in % of DMI).
17% - 4 000 kg milk/cow/year; 23% - 6 000 kg milk/cow/year; 27% - 8 000 kg milk/cow/year;
34% - 10 000 kg milk/cow/year; 38% - 12 000 kg milk/cow/year.

Figure 1 Energy intensity for feed supply in dairy farming in dependence on different milk yields without consideration of replacement rates

With higher milk yields this effect is less or contrary (>10 000 kg/cow/year). The increasing milk yields cannot compensate for the higher cumulative energy demand anymore, because of the increasing share of concentrates in the diet.

The influence of different replacement rates and milk yields on the energy intensity of feed supply in dairy farming is shown in Figure 2.

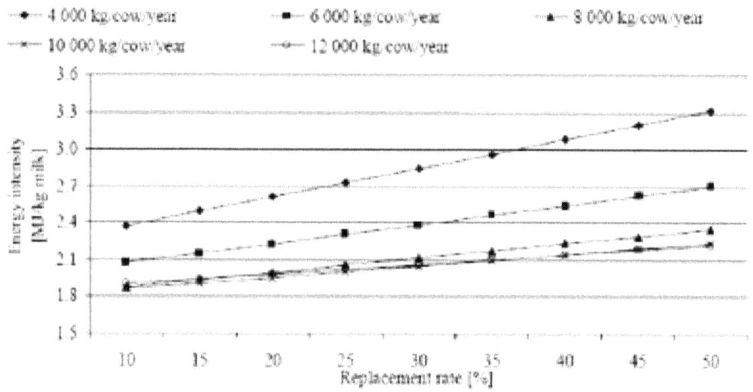

Figure 2 Energy intensity for feed supply in dependence on different replacement rates and milk yields.

These results indicate that with an increase in milk yield, the energy intensity for feed supply per cow decreases. However, for higher milk yields (8 000 kg/cow/year with a replacement rate of <30% or 10 000 kg/cow/year with a replacement rate of >30%) this effect decreases. The energy intensity for feed supply rises almost linearly with an increased replacement rate, irrespective of milk yield. The comparison of the different replacement rates and milk yields shows that both have a considerable influence on the energy intensity of feed supply. For example the energy intensity of feed supply for the production of one kg milk with an average milk yield of 6 000 kg/cow/year and a replacement rate of 20% is in the same range of that for a milk yield of 8 000 kg/cow/year and a replacement rate of 40% as well as that with a milk yield of 10 000 kg/cow/year and a replacement rate of 50%, respectively. Generally the replacement rate is increasing with an increase in milk yield.

Energy

Ruminants need a daily supply of all nutrients required for maintenance and production: milk, meat, growth and pregnancy. Quantitatively any type of nutrient can limit performance levels, but the most likely to be in short supply are energy and protein, this is especially true for high and average yielding cows. Both energy and protein must be considered. For energy, the feeding system uses the metabolisable energy (ME) in the feed as a basis to formulate rations. The ME is the energy remaining in the digested foodstuffs after the loss in faeces, urine, gases and body heat. The basic unit used to measure the energy content is the Megajoule (N4J).

Protein

The digestible crude protein (DCP) is widely used to evaluate protein requirements, and it corresponds to the crude protein that remains after losses in the faeces. However, a new system has been introduced which takes into account the degradability of the protein in the ration during digestion. It is a better system to calculate requirement levels, especially for high-yielding cows which have been shown to benefit from protein that escape microbial degradation in the rumen and is absorbed as amino-acids in the small intestine. Following this approach crude protein can be split into Rumen degradable Nitrogen (RDN) and Undegraded Dietary Nitrogen (UDN). Fish meal is for example considered as a good source of UDN.

Inorganic nitrogen sources from plants as well as other non-protein nitrogen, such as urea, are completely degraded by microbes in the rumen. Hence, the RDN is broken down by rumen microbes and used for their protein synthesis by the microbes. Later in the digestion process the microbes are themselves digested and the microbial protein becomes available to the animal Nevertheless this microbial synthesis is only optimal when the animal receives sufficient energy supplements. Therefore, if sufficient RDP is not available, the rate of digestion of fibrous as well as concentrate rich diets will be reduced. This leads to a reduction in intake, lower energy supply and reduced milk production.

On the other hand, some protein nitrogen can resist microbial breakdown in the rumen and can pass directly to the cows intestine. This feed protein fraction is called by-pass protein which is especially profitable for high-yielding cows. At a low level of productivity a cow can meet her protein requirements entirely from microbial protein and the diet only needs to contain degradable protein. This explains why such a cow can be fed with urea or chicken manure instead of high quality protein can meet the protein requirements. It is therefore important to have the optimum balance of UDP and RDP in the diet. Illustrates the increasing requirements for both as milk yield increases.

Minerals and Vitamins

The major mineral requirements for dairy cows are calcium and phosphorus. The calcium phosphorus ratio is important, and an imbalance can cause infertility. There are reserves of both elements in the skeleton. Pasture is often deficient in minerals, mainly in phosphorus. In some grazing areas, particularly uplands, dairy cows are subject to a mineral deficiency (especially daring lactation), thus obvious phosphorus deficiency symptoms are frequently observed (probably caused by very acid soil); these symptoms include sore legs, animals will limp, show long hooves and depraved appetite (pica). Seashells and fish meal are good mineral sources. Magnesium is frequently required to prevent hypomagnesaemia on grazing animals. All the required minerals can be successfully incorporated in feed blocks supplements.

Animals feeding on green pasture normally receive all vitamins that are needed. Vitamin supplements are often added to balanced rations prepared as feed to be used in animals housed and reared in intensive systems. Vitamin A is one of the most important vitamins in animal nutrition. This vitamin is found in the carotenoid pigment of green plants; Vitamins D, E and K are also present in green plants.

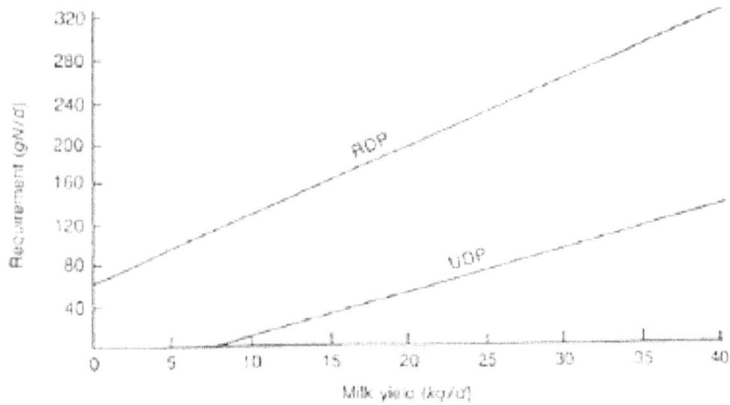

Figure 19. Protein requirements related to milk yield

SUMMARY

Cumulative energy demand for feed supply in dairy farming increases with an increase in milk yield and replacement rate. Whereas, energy intensity (cumulative energy demand/kg milk produced) decreases with increasing milk yields to a range of 8 000 to 10 000 kg/cow/year. Milk yields above this range effect only a slight increase of energy intensity.

8

Electricity Supply in a Dairy

Dairy farms today face challenges and opportunities fueled by rapidly rising energy costs and concerns about environmental impacts. Dairy farms use more energy than almost any other agricultural operation. Energy is used in the milking process, and for cooling and storing milk, heating water, lighting and ventilation. Determining the best energy efficiency and energy management opportunities for dairy farms will help reduce energy costs, enhance environmental quality and increase productivity and profitability.

Energy efficiency is often an inexpensive, quick and simple way to save money. This publication focuses on dairy equipment upgrades, new technologies and management practices for reducing energy consumption. Opportunities for cost savings and improved processes include the implementation of: variable speed drives for milk vacuum pumps and milk transfer systems, plate precoolers, heat recovery systems, energy-efficient light fixtures and efficient ventilation systems. Information on energy audits, financial incentives and further resources is included. The opportunities presented in this publication were developed for cow dairies but may be applicable to goat and sheep dairies as well.

MILKING PROCESS

The milking process consists of harvesting milk from the dairy cow and transporting the milk to a bulk tank for storage. This process can take place one or two or more times per day and most dairies schedule their lactations for a continuous milk supply throughout the year. On average, milking utilizes 18 percent of the electrical energy use on a dairy farm.

The vacuum pump operates during milking and equipment washing and is often the primary electrical energy user.

Vacuum pumps

The vacuum pump operates during the milk harvest and equipment washing and can consume 20 percent to 25 percent of all electrical energy use on a dairy farm. The vacuum pump creates negative air pressure that draws milk from the cow and assists in the milk flow from the milk receiver to either the bulk tank or the receiver bowl. The airflow also sends water and cleaning agents through the milking system during the washing process.

There are four main types of vacuum pumps: sliding vane rotary pumps, water-ring pumps, rotary lobe (blower) pumps and turbine pumps. Each of these operates and uses energy differently. Vacuum pumps become less efficient as the vacuum level increases (Ludington et al., 2004). Therefore, operating the vacuum pump at lower vacuum levels conserves energy.

Sizing the vacuum pump to meet the needs of the milking and washing system can reduce capital costs for equipment, reduce energy operating costs during the life cycle of the pump and ensure that the pump is performing properly. The size of the vacuum pump is dependent on all of the components in the system that admit air during operation. Pumps are generally sized by accounting for the amount of air admitted into the milking system during the milk harvest plus a 50 percent reserve to account for accidental air admittance and the wear and tear of parts (MMMC, 1993). Outdated industry standards used to recommend oversized vacuum pumps in part because of an incorrect belief that additional vacuum capacity is necessary for washing. An oversized vacuum pump may overheat as the pump ends its cycle. The current industry performance standard, ASAE Standard S518.2 Feb03, Milking Machine Installations, provides current sizing guidelines for dairy vacuum pumps (Ludington et al., 2004).

Variable-speed drives

A variable-speed drive (VSD), also referred to as an adjustable-speed drive or variable-frequency drive, is an energy-efficient technology used for controlling the vacuum level on sliding-vane rotary pumps and rotary-lobe pumps. Traditional vacuum systems run at maximum speed only, and rely on a vacuum regulator to vent excess air that is not being used during milking or washing operations. A VSD adjusts the speed of the pump motor to meet the vacuum demand so that equal amounts of air are being removed by the vacuum pump and entering the milking system. By eliminating the amount of air that would be admitted through a regulator, the motor uses less horsepower during most of the milking process. A VSD on a vacuum pump eliminates the need for a conventional regulator because less energy is delivered to the motor and its operating speeds are reduced. Energy operating costs of a vacuum system with a VSD can be reduced by up to 60 percent (Ludington et al., 2004). In addition, lowering the vacuum pump's RPMs can extend its life as there will be less wear and maintenance costs.

Vacuum pump maintenance

Following the manufacturer's maintenance schedule, along with routine maintenance on all vacuum pump systems, can save energy. Checking the temperature of the vacuum pump near the end of each milking is a good way to detect problems with the pump or electrical load. High motor temperatures may indicate the need for a vacuum adjustment, removal of exhaust restrictions, repair of rotary vane oiling systems or cleaning lobe blower pumps. Changes in the speed of the vacuum pump can be caused by air leaks in the milking system, degraded drain seals, loose pump belts, malfunctioning pulsators or a malfunctioning VSD vacuum sensor. A 5-cubic-foot-per-minute air leak in the vacuum system can amount to over $100 in wasted electrical costs (Peterson, 2008). Problems associated with vacuum pumps can be avoided by regularly checking vacuum levels and the operation of the VSD.

Milk cooling systems

Cooling milk accounts for most of the electrical energy consumption on a dairy farm. Milk is harvested from a cow at around 95 to 99 degrees Fahrenheit (F). To maintain high milk quality, including low bacteria counts, the raw milk temperature needs to be lowered quickly to around 38 degrees. Refrigeration systems consist of a bulk tank, a motor-driven compressor unit, an evaporator and a condenser unit. The system operates by pumping a refrigeration coolant in a cycle that gives off heat as it condenses. Heat exchangers cooled by well water, variable-speed drives on the milk pump, refrigeration heat recovery units and scroll compressors are all energy conservation technologies that can reduce the energy consumed in the cooling system.

There are two types of milk cooling systems used on dairy farms. A direct-expansion cooling system cools the milk directly in the bulk tank by using evaporator plates that expand in the lower portion of the tank. Indirect, or instant, cooling systems rely on heat exchangers to precool the milk before it enters the bulk tank. A precooling heat exchanger can lower milk temperatures as much as 40 degrees, resulting in refrigeration energy reductions of about 60 percent (Sanford, 2003b).

Heat exchangers

Heat exchangers used for precooling raw milk transfer the heat from the milk to an intermediate cooling fluid, such as well water or a water-glycol solution, by having the milk and coolant flow in concurrent or counterflow configurations. In a concurrent flow configuration, the milk and coolant flow in the same direction. The fluids flow in opposite directions in a counterflow pattern. Counterflow systems allow for lower milk temperatures. The amount of cooling that takes place depends on the flow of the milk in relation to the flow of the coolant, in addition to the heat transfer area, the number of times the milk passes through the coolant and the temperature of the coolant.

Well-water heat exchangers have been implemented in precooling systems for more than 20 years and can reduce cooling costs by 0.2 to 0.3 kWh/cwt (hundred weight of milk) (Ludington et al., 2004). There are two types of heat exchangers used for precooling: shell-and-tube heat exchangers and plate heat exchangers. Shell-and-tube heat exchangers consist of a single small tube or multiple small tubes inside of a larger tube. The milk flows through the smaller tube(s) while the well water flows through the larger tube. Plate heat exchangers are made up of a series of metal plates that are located side by side. Milk and well water run in opposite flow channels and the well water absorbs the heat from the milk on the opposite side of each plate. Plate heat exchangers are relatively small in size and, unlike shell-and-tube heat exchangers, additional plates can be added to increase capacity.

Heat exchangers are designed so that the milk and coolant make single or multiple passes between the plates or tubes. In a single-pass heat exchanger, a circuit is arranged so that the two fluids are in contact as they go through separate sets of plates or tubes before exiting the heat exchanger. Multiple-pass heat exchangers route the milk and coolant flow through two or more circuits, putting the milk channels in contact with the coolant channels for longer periods of time. A heat exchanger can be designed so the milk travels through multiple circuits and the coolant flows through a single pass, or circuit. Farms that milk for long periods of time can use heat exchangers that are designed for two coolants. The use of two different cooling fluids, such as first precooling with well water and then with chilled water or a glycol-water solution, can cool the milk as it is harvested.

Precooler energy savings

Installing a properly sized precooler can reduce refrigeration energy consumption by about 60 percent (Sanford, 2003b). A properly sized well water heat exchanger can reduce milk temperatures to within 5 to 10 degrees of the groundwater temperature. Higher efficiency for precooling can be achieved with a 1-1 ratio of coolant to milk flow and by using the largest water lines possible in order to maximize the coolant flow. Other factors that determine the energy and economic savings of a precooler include herd size, number and size of compressors, type of coolant used and the age of the bulk tank. An added benefit of using water for precooling is that it can afterwards be used for livestock drinking water. Cows depend on drinking water to produce milk and prefer warm water to cold. The warm water can also be used to perform other tasks, such as washing and heating.

A well water precooler (far right) cuts compressor use by using groundwater to pre-cool the milk.

VARIABLE-SPEED MILK PUMPS

As previously stated, a vacuum is required to harvest milk. The milk, however, cannot transfer from a pipeline that is under a vacuum to a bulk tank that is at normal atmospheric pressure. To compensate for the difference in pressure, milk can flow into a receiver bowl that triggers a milk transfer pump to push the milk from the receiver bowl either through a heat exchanger or directly into the bulk tank. Another option that eliminates the receiver bowl and milk transfer pump requires putting the bulk tank under a vacuum by placing rubber seals on the bulk tank covers.

A variable-speed milk transfer pump can further reduce energy use by slowing the flow rate of milk through the heat exchanger. A lower and more continuous milk flow rate through the heat exchanger increases the coolant-to-milk ratio and results in greater milk cooling. Milk can be cooled by an additional 15 to 20 degrees by installing a variable-speed milk transfer pump (Sanford, 2004c).

Refrigeration heat recovery units

Refrigeration heat recovery units (RHR) capture the waste refrigeration heat from the condenser to preheat water before it is transferred to a water heater. An RHR unit can recover 20 to 60 percent of the energy that is required to cool milk for storage (Sanford, 2004b). Depending on the number of cows being milked, the RHR storage tank should be sized to provide enough hot water for one milking. This includes hot water for washing the milking system and for washing the bulk tank. The payback for purchasing an RHR unit depends on the savings from heating water and will vary greatly from farm to farm.

Types of refrigeration heat recovery units

There are two types of RHR units: desuperheating and fully condensing units. A desuperheating unit operates only when the refrigeration system operates and produces water temperatures around 95 to 115 degrees (Ludington et al., 2004). These units can be inexpensive to add, but only part of the available heat can be transferred to the water. Nearly all of the available refrigeration heat is recovered as hot water from a fully condensing unit, with temperatures reaching 120 to 140 degrees (Ludington et al., 2004). This type of RHR unit meets the requirements of the cooling system by using a valve to control the flow of cold water through the heat exchanger. Once the storage tank is full, excess warm water can continue to be used and not wasted. A fully condensing RHR unit replaces the need for a regular condenser in the refrigeration system and is best integrated as part of a new or replacement refrigeration system. If nearly all of the energy captured by the RHR unit is used for preheating water, then a precooler is not needed to remove heat from the milk and may, in fact, increase overall energy costs.

Compressors

The function of a compressor is to compress the cold, low-pressure refrigerant gas from the evaporator to a high-pressure, high-temperature state for condensing. The three types of refrigeration compressors used for milk cooling on dairy farms are the reciprocating, scroll and discus. The reciprocating compressor is the most common and can be either open type, hermetic or accessible hermetic. The scroll and discus compressors are newer and more efficient than reciprocating compressors.

Research has shown that replacing a failed reciprocating compressor with a new scroll compressor can reduce milk cooling costs by 20 percent because of a reduction in the electrical demand (Ludington et al., 2004).

Heating water

Hot water is essential for producing high-quality milk on dairy farms and is primarily used for cleaning milking systems. The amount of hot water required varies from farm to farm and depends on the size of the milking herd and the type and size of the milking system. On average, a minimum of 4 gallons of water at 170 degrees are required for each milking unit (Ludington et al., 2004). Hot water is generally produced by water heaters that use either fuel oil, natural or propane gas, solar energy or electricity. For some dairy farms, heating water can account for about 25 percent of the total energy used on the farm (Sanford, 2003a).

Types of water heaters

There are several ways to reduce the amount of energy used for heating water. Whether using a direct or indirect water heater, overall efficiency is determined by the combustion efficiency of the fuel source and the amount of heat loss from the storage tank, known as standby loss. A direct water heater combines the water storage tank and the heating element. The storage tank in an indirect water heater contains a heat exchanger that is connected to a separate boiler unit. Insulating the storage tank and connecting pipes can reduce standby losses for both types of water heaters. The sides and top of an electric water heater and the sides of gas and oil water heaters can be insulated.

Water temperatures

Proper water temperatures are essential for cleaning milking equipment. Using warm or cold water when possible can help reduce energy costs. A warm-water pre-rinse, about 95 to 110 degrees, does not cause the milk to stick to the equipment, which is a problem with cold water. Additionally, a warm water rinse does not cause mineral and protein deposits in the pipeline, as is often the case when using hot water. The wash cycle requires water temperatures between 120 and 170 degrees. Dairy farms require the use of commercial water heaters since residential electric water heaters are thermostatically controlled not to go above 140 degrees because of federal safety regulations. The thermostat range for commercial water heaters is 100 to 180 degrees. A cold acid rinse will control mineral and bacterial buildup in the milk lines and will help conserve hot water. There are many types of dairy acid sanitizers and each one may have different rinse instructions. Most of these sanitizers are not listed on the National List of Allowed and Prohibited Substances for use in certified organic production and processing (National Organic Program) but some may be permitted for use on organic dairies (USDA, 2000). Check with your organic certifier to see if they have any restrictions for use of dairy acid sanitizers.

Electric control devices and peak demands

Using a control device for electric water heaters can reduce electricity use during periods of expected increases in demand above the average supply level, also known as peak demand. Time clocks provide a simple and cost-effective way to control water heaters. Shifting electrical loads to off-peak or low demand hours and off-peak utility rates not only ensures farmers of having hot water at a reasonable cost, but also benefits the utility provider by allowing direct control of the energy demand on their system. Many utility companies offer incentive rates or rebates to encourage farmers to participate in electric water heater control programs. Not all utility companies offer off-peak utility rates, so it is best to check with local service providers to see if off-peak rates are available.

Lighting

Appropriate lighting can improve productivity and safety on a dairy farm. On average, lighting represents 17 percent of total dairy farm electrical energy use (Peterson, 2008). Optimal lighting conditions may increase milk productivity and conserve energy consumption. Factors that contribute to increased milk production include the type of light, the amount of light provided per watt, the temperature of the work area, the height of the ceilings and the length of the lighting period.

Lighting requirements on a dairy farm can be divided into three categories. The first category is visually intensive task lighting, which requires the highest lighting levels (Ludington et al., 2004). Areas that benefit from this type of lighting include milking parlors; equipment washing, equipment maintenance and repair areas; offices; maternity and veterinary treatment areas; and utility rooms. The second category includes lighting for holding areas, feeding areas, animal sorting and observation and general cleanup. These areas and tasks require high to moderate lighting levels. Finally, low to moderate lighting levels are adequate for general lighting for livestock resting areas, passageway lighting, general room lighting and indoor and outdoor security lighting.

Lamp types

There are three types of lighting systems used on dairy farms: incandescent lamps, fluorescent lamps and high-intensity discharge lamps.

Incandescent lamps

Incandescent lamps are common on many dairy farms, although today they are considered to be an inefficient light source. Their light output is only 10 percent, with the remaining 90 percent of energy given off as heat (Hiatt, 2008).

Fluorescent lights

Fluorescent lights are made of a frame, ballast, light socket and the lamp(s). Ballasts provide the needed voltage, current and waveform for starting and operating fluorescent and high-intensity discharge lamps. The lamps are available in many different lengths and diameters, shapes, color temperature ratings and minimum operating temperature ranges. Fluorescent lamps range in length from 6 to 96 inches long. The diameter of the lamp is expressed in eighths of an inch and lamps come in the following diameters: T-5, T-6, T-8, T-9, T-10, T-12, and T-17. A T-12 lamp is 12/8 inch, or 1.5 inches, in diameter, while a T-8 is 8/8 inch, or 1 inch in diameter (Sanford, 2004a).

Fluorescent lamps are linear (straight), circular or U-shaped. The color temperature index, or CCT, refers to the color appearance of the light emitted by a lamp (Sanford, 2004a). It is indicated in degrees Kelvin (K). Lower temperature values produce a red or orange color and refer to a warm emitted light. Higher temperature values produce a blue, or cool, light source. The CCT values for fluorescent lamps range from 3,000 to 5,000 degrees K, while the average CCT value of an incandescent lamp is around 2,800 degrees K. Minimum operating temperatures for fluorescent lamps are between -20 and 50 degrees Fahrenheit, depending on whether the ballast is electromagnetic or the more energy-efficient electric type.

The most common type of fluorescent lamp is the T-12. However, smaller diameter lamps provide more light for a watt of power. Compared to a T-12 lamp and ballast, the T-8 fluorescent lamp provides about 15 percent more light, or lumens, and the ballasts are 40 percent more efficient (Sanford, 2004a). The most energy-efficient fluorescent lamp is the T-5, but that lamp tends to produce too much heat in an enclosed fixture. Enclosing fixtures in agriculture lighting systems protects the housing from corrosion and protects the lamp from dust, humidity and manure gases, all of which can shorten the life of a lamp.

COMPACT FLUORESCENT LAMPS

A compact fluorescent lamp (CFL) uses about 75 percent less energy, measured in watts, for the same amount of light as an incandescent bulb. CFLs are directly interchangeable with incandescent lamps, so changing a 60-watt incandescent lamp to a 20-watt CFL cuts energy use by one third. CFLs also last six to 10 times longer than incandescent light sources (Sanford, 2004a). CFLs can be used outdoors, but should be covered to protect the lamps from all weather conditions. Low temperatures, both indoors and outdoors, may reduce light levels. In order to reach full output at cooler temperatures, CFLs require a warm-up period.

High-intensity discharge lights

High intensity discharge (HID) lamps are a very bright and efficient light source. HID lamps often require a warm-up period and generate light by heating a gas mixture in the lamp. HID lamps must be mounted high in order to spread the light and to avoid a glare. Some HID fixtures can be equipped with a diffuser that distributes the light horizontally. The four common types of HID lights are mercury vapor, metal halide, high-pressure sodium and low-pressure sodium.

Mercury Vapor Lamps

Mercury vapor lamps emit a favorable blue-white color that mimics daylight. However, mercury vapor lamps are the least efficient of the HID lamps, emitting only 32 lumens for a watt. In addition, their disposal poses an environmental risk due to the mercury gas.

Metal Halide Lamps

Metal halide HID lamps provide the best color of all the HID lamps and emit about 60 lumens a watt, making them twice as efficient as a mercury vapor lamp. A newer pulse-start lamp technology is available for metal halide lamps. The pulse-start system can extend lamp life by half over the standard metal halide lamp, provide about 8 percent more lumens per watt and allow for faster warmups and restarts (Sanford, 2004a). Pulse-start metal halide lamps use a different type of ballast and are not interchangeable with standard metal halide lamps.

High- And Low-pressure Sodium Lamps

Slightly more efficient than metal halide lamps are high- and low-pressure sodium lamps. High-pressure sodium lamps emit about 95 lumens a watt and give off an orange light. A 100-watt high-pressure sodium lamp can provide 2.5 times more light than a 100-watt mercury vapor lamp while using significantly less energy. High-pressure sodium lamps are often used for mixed outdoor and indoor lighting applications. Low-pressure sodium lamps are the most efficient HID light source, but produce a poor color of light. Low-pressure sodium lamps are often used for outdoor security lighting.

Lighting Sodium efficiency measures

Replacing inefficient light sources with an appropriate and higher-efficiency light can result in better task lighting with energy savings that continue over the life of the lamp. Energy conservation opportunities involve changing incandescent lamps to compact fluorescents, upgrading to smaller diameter and more efficient fluorescent lamps and upgrading to HID lighting. Since lighting fixtures distribute light in different patterns, the barn layout and use should be considered in any lighting decision.

Energy savings can be achieved by using simple and inexpensive technologies that go beyond lamp replacement. Some technologies incorporate programmable logic controllers and other computer-based control systems and include timers, dusk-to-dawn photo controllers, half-night controllers and motion detectors. New and more efficient technologies are continually being developed.

Factors affecting lamp life

Several factors can shorten the life of a lamp. These include:

- Level of manure gases
- Humidity

- Dust
- Frequent switching on and off

Long-day lighting

Long-day lighting is a new development for dairy farms that involves increasing milk production by extending the day length with artificial light. According to researchers, dairy cows are day-length sensitive and can be stimulated to increase milk production by providing 16 to 18 hours of uninterrupted lighting followed by 6 to 8 hours of darkness (Sanford, 2004a). By implementing an efficient long-day lighting system, milk production can be increased bv 5 to 16 percent.

Ventilation

Proper ventilation is needed in dairy barns throughout the year to help maintain animal health and productivity, the barn's structural integrity, milk quality and a comfortable work environment for the laborers. The air inside a barn becomes warm and humid as cows continuously produce heat and moisture. An efficient ventilation system brings fresh air that is cooler and drier into the barn, mixes and circulates it, and exhausts and dilutes moisture, dust, manure gases and odors, and airborne contaminates.

Cows that are exposed to fresh air are less likely to develop heat stress, respiratory ailments, mastitis, reproductive problems and other diseases that could decrease milk production. Many dairy producers are realizing the inherent benefits of a grass-based product ion system. Grass-based dairies utilize an ecological approach to health care by relying on natural immunity that comes with pasture access. For more information on grass-based dairying, see the ATTRA publication Dairy Production on Pasture: An Introduction to Grass-Based and Seasonal Dairying. According to experts, heat stress in dairy cows begins when surrounding temperatures exceed 65 degrees and the relative humidity is 40 percent or higher (Ludington et al., 2004). Maintaining a healthy atmosphere by controlling mold, spores, dust and other irritants also helps reduce the chances of milk contamination and lowers health risks to the farmers. Good ventilation also reduces building deterioration due to excess moisture, which can cause rot, mildew and electrical problems.

Types of ventilation

There are two types of ventilation: natural and mechanical. Natural ventilation uses the least amount of energy and requires the exchange of air, the ability to control ventilation rates, flexibility to provide the cows a comfortable environment throughout the year and good barn construction (Wood Gay, 2002). Barns that are in the path of prevailing winds and designed with adjustable side and end walls can take advantage of breezes and thermal buoyancy and minimize unexpected air exchanges.

Mechanically ventilated barns require air inlets and exhaust fans opposite one another. Exhaust fans are powered by an electric motor and contain hoods, protective screens and louvers or shutters. Each fan targets a certain area to provide uniform control of moisture and gases (Peterson, 2008). Ventilation fans used in livestock buildings are typically axial flow fans, which are designed to let air pass straight through the fan parallel to the fan blade drive shaft (Arnold and Veenhuizen, 1994). Traditional ventilation systems are designed with air inlets and exhaust fans placed on opposite ends of a building, whereas air inlets and exhaust fans in a cross-ventilation system are located on opposite sides of a building.

Fan selection

A properly designed and sized ventilation fan system should provide the required amounts of ventilation needed to maintain a comfortable environment for the cows and the laborers during each season. The blade size and shape, fan speed in terms of revolutions per minute (rpm), motor horsepower and housing design determine how much air a fan can move. This is known as fan capacity and is measured cubic feet per minute (cfm). The size of ventilation system needed depends on its capacity to remove excess moisture during the milder and colder months of the year while forcing cool air through the building during the summer to reduce animal heat stress. Exhaust fans range in size from 5 inches in diameter to 72 inches. Exhaust fan motors may be as small as 1/20 horsepower and may exceed more than 50 horsepower (NFEC).

Static pressure is the difference in pressure between inside and outside of a building, ventilation fan or air inlet (Ludington et al., 2004). It is measured in inches of water or inches of mercury. The static pressure of an efficient exhaust fan should be between 0.05 and .125 inches of water and is measured with a manometer.

Circulation fans

Circulation fans are mounted inside buildings to help improve air movement and to reduce dead air spots. There are different types of circulation fans, all of which have propeller blades attached directly to a motor or to a motor-and-belt drive system. Two common types of circulator fans used in dairy housing systems are box fans and high-volume low-speed fans (HVLS). Box-type fans generally operate at moderate to high levels of efficiency, depending on the fan diameter, with the larger diameters being more efficient (Sanford, 2004d).

HVLS fans are large-diameter circulation fans with diameters ranging from 8 to 24 feet. One 24-foot HVLS fan can move the same amount of air as six 48-inch high-speed box fans and can save up to 3.3 kilowatts in an hour of operation (Sanford, 2004d). HVLS fans save energy by replacing existing circulation fans that are not efficient. In addition, HVLS fans tend to be quiet, reduce moisture in buildings and reduce fly and bird populations in barns.

Fan ratings

The Air Movement and Control Association International (AMCA) is an independent testing laboratory that rates fans on their ability to move air. Fans that meet certain AMCA or other reliable testing performance criteria receive a certified rating seal of approval. This rating can be helpful to select the proper fan for an intended use. For more information on the AMCA, visit their Web site at www.amca.org. The Bioenvironmental and Structural System (BESS) Laboratory provides research, product testing and education on agricultural fans. Performance test results are available on their Web site at www.bess.uiuc.edu.

Ventilation control systems

The use of thermostats to control ventilation systems reduces energy consumption by having the fans operate only when needed. Properly located thermostats should accurately reflect the average temperature inside the building. Mounting a thermostat at least 10 feet from a fan and at least 1 foot below the ceiling reduces the thermostat's exposure to adverse factors, such as sunlight and air intake drafts, that can alter the thermostat's temperature reading (Wells, 1990). All control systems should be located so they cannot be bumped by the livestock.

Timer controls and moisture sensors can also be used to control ventilation rates. However, the use of these control devices may not save energy. Timer controls cycle motors on and off to meet set ventilation rates and may use more energy to start the motors from the off mode. Timer controls also create larger temperature swings. Furthermore, moisture sensors have been reported not to work adequately in barns (Wells, 1990).

Ventilation system maintenance

Dairy ventilation systems require routine maintenance to keep fans operating at high performance levels. Poorly maintained fans and obstruction to air inlets and fan outlets can reduce fan efficiency by as much as 40 to 50 percent. Cleaning fan parts, lubricating bearings and other moving parts, checking belt tension and alignment and removing any obstructions will keep fans performing at peak efficiency and reduce energy costs. Long-term efficiency can be improved by repairing or replacing bent, damaged or misaligned fan blades, shutters, guards and other fan parts. An air velocity meter can be used to detect reductions in the performance of the ventilation system. Control systems should be cleaned regularly and thermostats should be checked yearly for accuracy by comparing their readings against a mercury thermometer.

Factors that affect the efficiency of a livestock ventilation system

- Ventilation system design
- Fan blade type and design
- Efficiency of electric motors to power fans
- Use of fan control system
- Routine maintenance and cleaning

SUMMARY

This chapter provides an examination of the dairy industry, which of all the land extensive agricultural industries is the most dependent upon the consistent reliability of electricity provision and is also a much greater relative user of electricity as an input into production than the other land extensive or broadacre industries.

Electricity usage for agricultural production purposes is on a daily basis. Major requirements are for:

- The milking machinery, which is used a minimum of twice and sometimes three times per day
- The chillers for the raw milk before it is picked up
- Refrigerators for animal health products, such as vaccines
- Hot water systems for cleaning equipment in the dairy
- Computers and other electronic data and sensory systems for quality control and recording purposes
- Irrigation pumps for some farms and at various times
- Farm workshop requirements

9

Equipment and Materials

The basic layout of the three main types of milking machines are the same. Each has a pump to remove air from the vacuum pipeline, a vacuum regulator and a container to collect the milk that comes into the teatcup assembly during milking.

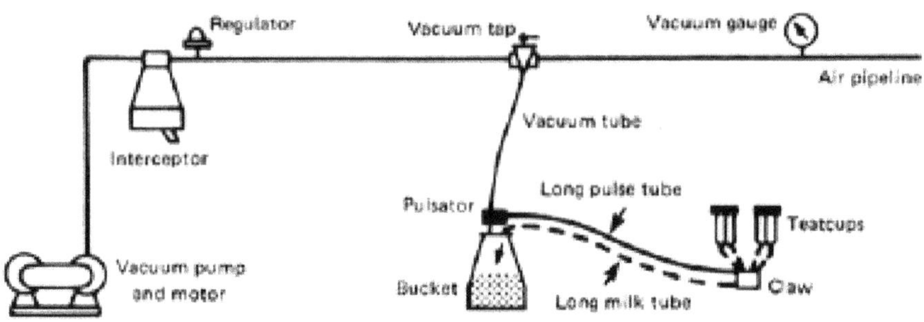

(a) Bucket (cowshed)

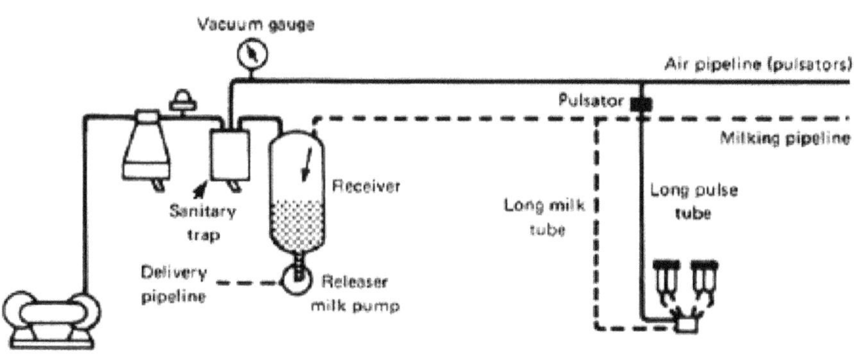

(b) Milking pipeline (cowshed and parlour)

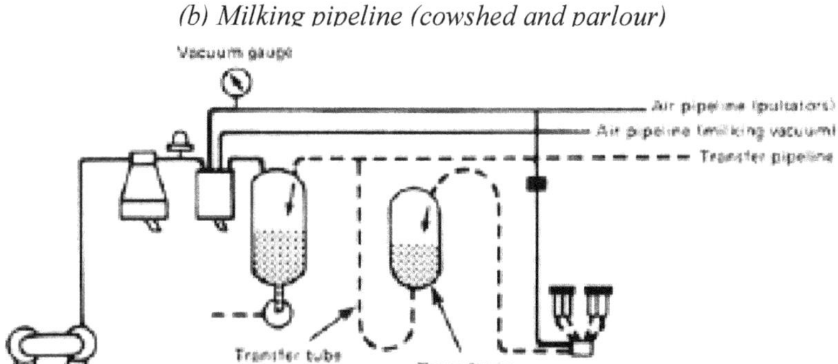

(c) Recorder (parlour)

Principal types of milking machines

Milk - - -
Air ———

MILKING MACHINES AND EQUIPMENT

Basic construction

The principle of machine milking is to extract milk from the cow by vacuum. The machines are designed to apply a constant vacuum to the end of the teat to suck the milk out and convey it to a suitable container, and to give a periodic squeeze applied externally to the whole of the teat to maintain blood circulation.

A milking machine installation consists of a pipework system linking various vessels and other components which together provide the flow paths for air and milk. The forces necessary to move air and milk through the system arise from the fact that it is maintained at a vacuum. Thus it is atmospheric pressure which forces air, and intra-mammary milk pressure which forces milk, into the system and the combination of these forces causes flow. To be a continuous operation it is necessary to remove air and milk from the system at appropriate rates.

Although milking machines have now developed into systems that show considerable diversity they have the same basic components. The air is removed by a vacuum pump at a constant rate. In a bucket or direct-to-can machine milk is removed from the system by disconnecting the milk container; in milking pipeline and recorder machines the milk is removed by a milk pump or releaser.

Fig. shows the flow of air and milk through three basic types of machine during normal milking. In the bucket (or direct-to-can) machine the milk enters the teatcups and travels through the short milk tubes to the claw where air is admitted and the milk and air travel along the long milk tube to the bucket (or can). The milk remains in the bucket (or can) and the air separates to pass up the vacuum tube to the vacuum pipeline. The pulsator which is usually fixed on the bucket lid admits air intermittently and this passes along the long pulse tube to the teatcup chambers. To control the vacuum at a predetermined level air is also admitted to the system through a vacuum regulator which is fitted on the vacuum pipeline near to the milking points.

Vacuum And Milk Flow

When the milk from the claw is raised to a pipeline this can markedly reduce the vacuum at the teat because of the weight of milk in the long milk tube. The reduction in vacuum can be much reduced by bleeding air through a small hole in the clawpiece

In addition to the designed sources of air admission, air can be drawn into the teatcups past the teat and also when a milk container is changed or emptied. In a poorly maintained machine there may also be inward leakage of air at joints or points of damage. To maintain the working vacuum the vacuum pump extracts the air admitted into the system by compressing it so that it can be discharged to atmosphere.

In pipeline milking machines the flow pattern is similar to the bucket machine except that milk and air from each claw flow either directly to arecorder vessel where air and milk are separated, and/or through the milking pipeline to a common receiver vessel where milk and air are separated. There is no further air admission at this point when a motor driven releaser milk pump is used to empty the receiver. Other types of releaser (eg. pulsator controlled spit chamber and double chambered weight operated) admit air.

Where air and milk are transported together the flow pattern becomes complex depending on various factors particularly the volume of air relative to milk or air:milk ratio. Air is normally admitted in to the claw at a rate of 4 to 8 l/min. A milk flowrate for a fast milking cow will be about 6 l/min, giving an air:milk ratio of 0.7:1 to 1.2:1. Towards the end of milking when the milk flowrate has decreased to 0.25 l/min the ratio becomes 16:1 to 32:1.

The air:milk ratio becomes important where milk has to be elevated from the claw as in milk pipeline and recorder machines other than those with low level milk pipelines. Elevating a liquid, as distinct from a gas, involves a loss of potential energy and this is compensated for by a change of vacuum. Thus elevating a column of milk in a vacuum system through 1 m height reduces the vacuum by about 10 kpa. Therefore if the vacuum at the top of the column is 51 kpa it will be only 41 kpa t the bottom. This vacuum drop is markedly reduced by the admixture of air. If the air:milk ratio is 1:1 the weight of milk in the column is halved and the vacuum drop becomes only 5 kpa; if it is 9:1 the vacuum drop is only 1 kpa.

Under vacuum liquids cannot flow against gravity (ie. uphill) except as a column which fills the bore of the tube. Where the tube contains air and milk the liquid forms plugs which are separated by pockets of air in the proportion determined by the air:milk ratio.

VACUUM MEASUREMENT

Vacuum is a pressure below atmospheric pressure.

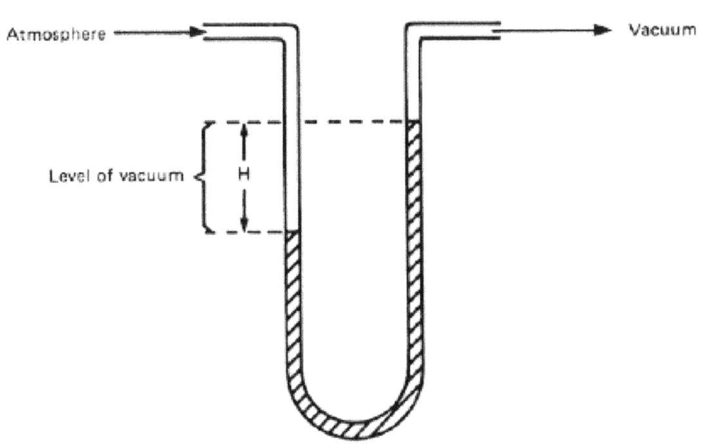

It can be measured as pressure difference with a mercury manometer in mm Hg. The standard pressure is now kilopascals (kPa) with 100 kPa equal to the pressure difference between atmospheric pressure and absolute vacuum.

Measurement Of Vacuum

Vacuum is a pressure below atmospheric pressure, the term "negative pressure" is sometimes used but in milking machine terms it may be considered to mean "vacuum" measured on a scale in which atmospheric pressure at the time and place of measurement is zero vacuum.

Vacuum can be measured in a variety of units. A commonly used measure is the linear difference in height between two columns of mercury in a 'U' tube when one of the columns of mercury is subjected to a vacuum and the other open to atmosphere. (see diagram of 'U' tube). The difference in height of the levels is supported by atmospheric pressure.

In the past the most commonly used units have been inches, millimetres or centimetres of mercury (in Hg, mmHg or cmHg). Units now adopted by the International Standards Organisation (ISO) for International Standards of milking machines for vacuum measurement are kilopascals (kPa) with zero (0) kPa being equal to atmospheric pressure and 100 kPa absolute vacuum. Equivalent relationships for values of vacuum levels are:-

1 mmHg = 0.133 kPa

1 inHg = 3.386 kPa

Equivalents for vacuum levels of 50 kPa and 44 kPa that are the most commonly used levels for milking cows:

50 kPa = 14.8 inHg = 375 mmHg

44 kPa = 13.0 inHg = 330 mmHg

Action of the Milking Machine in Each Pulsation Cycle

Pulsator connects pulsation chamber to vacuum, liner opens and milk flows.

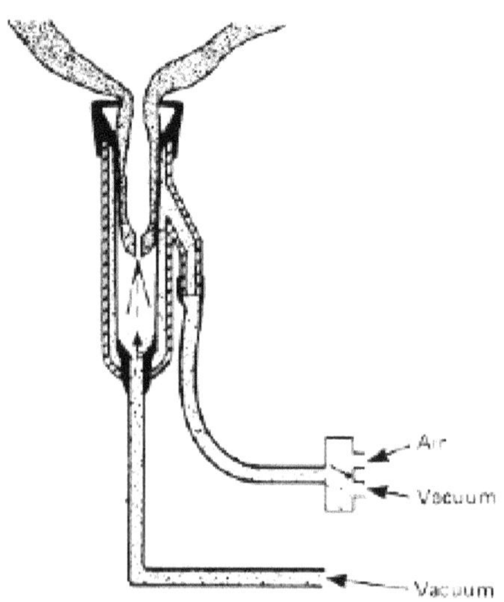

Pulsator connects pulsation chamber to atmosphere, liner collapses, squeezes the teat duct and prevents milk flow.

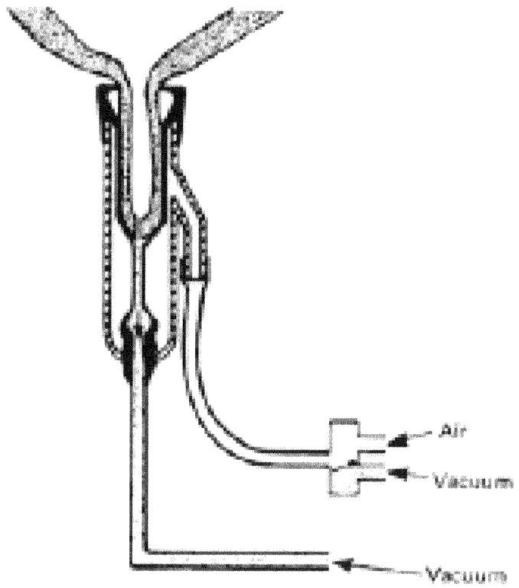

Milking Rate

Milk flow from the teats increase with:

- increasing vacuum but strippings also inOcrease.
  (Normal vacuum range 40-50 kPa).
- increasing pulsation rate but this increases air to be pumped from the machine.
  (Normal rates 50-60 cycles/min)
- widening pulsation ratio, ie liner open to liner collapse time.
  (Normal range 50/50 to 70/30)

Teatcup Liners

Teatcup liners have important effects on milk flow and completion of milking.

- the liner mouthpiece affects the quantity of strippings
- narrow bore (<24 mm) and low tension (stretch) milk more slowly.
- choose liners appropriate for the size of cows teats to be milked.

Action of the milking machine

The principles of machine milking were established many years ago and the basic method described below, is used in virtually all commercial milking machines although in a minority some modifications are made. The teatcup liner is the only equipment that comes into contact with the cows teats. The continuous vacuum within the liner causes the teat duct (streak canal) to open and the milk to flow because of the pressure difference between the milk in the teat and vacuum. To prevent damage or pain to the teat that would be caused by the continuous vacuum a system called pulsation is used. This makes the liners collapse on and below the teats about once each second massaging the teat and maintaining a more normal blood flow. In each pulsation cycle milk does not flow from the teat when the collapsed liner squeezes the teat duct.

Providing the cows 'let down' (ejection) has occurred the flow rate from the teat depends largely on the bore of the teat duct which is an inherent factor and not subject to management practices or training. Flow rates are also influenced by the mechanical properties of the milking machine. After the teat cups have been attached the flow rate reaches a maximum in about one minute, usually within the range of 2-5 kg/minute and the total milk flow period will range from 2 to about 8 minutes depending upon milk yield. Flow rates decline at the end of milking and when flow ceases there is usually a small amount of milk trapped in the sinus of the udder which can be removed by pulling ownwards on the clawpiece and massaging the udder (ie. machine stripping). With modern designs of liner the quantity of strippings is small (ie. less than 0.3 kg) and machine stripping is not usually practiced. The small amounts of milk that are left do not affect milk yield or the average chemical composition of the milk obtained or mastitis.

Vacuum Fluctuations

The fluctuations in vacuum in the teatcup liner have important effects on mastitis and milk flow. There are two types.

Irregular fluctuations

These occur when the teatcup liners slip or fall from the teats or air enters when milking units are changed carelessly. Vacuum recovery is slow if there is inadequate vacuum pump capacity.

Cyclic (regular) fluctuations

The cyclic movements of the liner in each pulsation cycle increase and decrease the volume of the liner under the teat. When milk is flowing this can cause marked changes in vacuum below the teat. This can be reduced by:

- using wide bore short milk tubes (>8 mm)
- ensuring claw or short milk tube air bleeds are not blocked.

The main milking machine factor affecting milk flow rate is the liner vacuum. Raising vacuum levels gives faster milking but also increases strip yields and in practice a compromise level of about half atmospheric pressure is used (ie. 40-50 kPa, 300-375 mm/Hg). The pulsation characteristics also affect flow. An increased pulsation frequency (rate) gives faster milking but because this greatly increases the air admission in the machine and therefore the required pump capacity it is usual to keep pulsation rates at 50-60 cycles of liner opening and closing per minute. Because milk flow ceases in each pulsation cycle when the liner is collapsed on the teat, faster flow rates are obtained by using a wider pulsation ratio (ie. ratio of liner open time to liner collapsed time). For udder health reasons the ratios are usually not greater than 70:30. The design of the liner can also affect the flow rate but modern liners tend to have similar flow properties. Narrow bore (<24 mm) liners and those with low tension in the barrel (ie. not stretched in the teat cup) milk more slowly. The most important characteristic of the performance of a liner is the amount of strippings left at the end of milking which is mainly determined by the dimensions and hardness of the mouthpiece. Liner design is largely empirical and farmers determine the best liners for minimum strippings by trial and error.

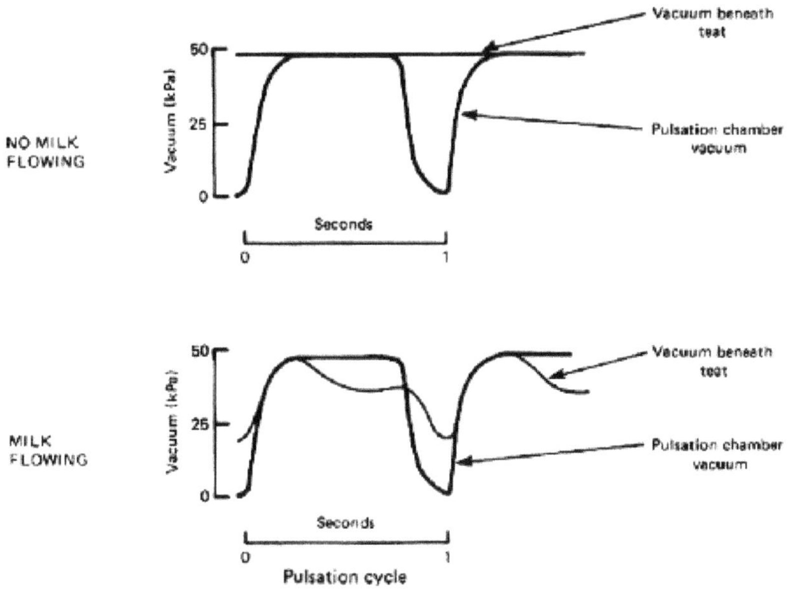

Although the teat cup liners are connected to a pipeline maintained at a constant vacuum level there can be considerable vacuum fluctuations in the liners, mainly due to the movements of the liner wall brought about by pulsation. When the liner is opening and the milk is moving away from the teat along the short milk tubes the vacuum below the teat will increase markedly due to the increased volume of the liner and the kinetic energy of the milk in transit. These variations in vacuum occur with each pulsation cycle and are called 'cyclic fluctuations'. The fluctuations are increased with adventitious air admission that occurs when liners slip on the teats or when machines are removed from adjacent cows udders (irregular fluctuations). These cyclic and irregular fluctuations generate impact forces which are important factors causing mastitis. Various methods have been used to reduce fluctuations to prevent them having deleterious effects. The most important are the provision of adequate airbleed holes in the claw or short milk tubes to aid milk flow and prevent flooding in the liner. Also useful has been an increase in the internal diameter of short milk tubes (eg. greater than 8 mm). These can be reduced by other modifications. In some designs the basic system of machine milking has also been modified to give lower levels of vacuum at the start and end of milking when there is no milk flow or by the inclusion of a positive pressure phase in each pulsation cycle to give increased let down stimulation.

VACUUM PUMP

Most vacuum pumps are rotary oil sealed positive displacement pumps Air is drawn in through the inlet port by the rotation of the rotor, compressed and ejected to atmosphere.

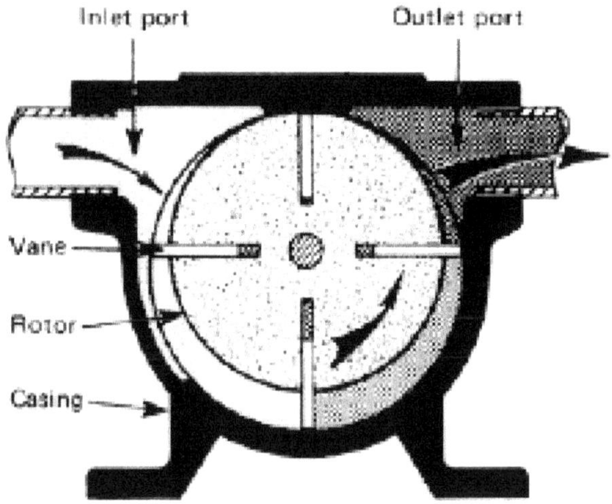

INTERCEPTOR

A trap fitted in the main vacuum line to prevent liquid and dust being sucked into the vacuum pump

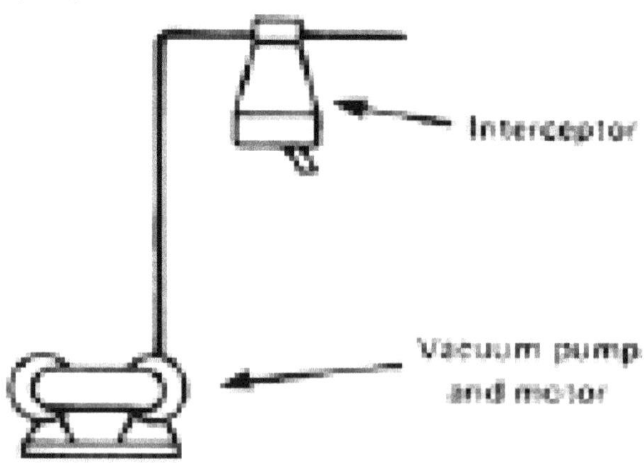

Components

VACUUM SYSTEM

The function of the vacuum pump is to extract air from the pipeline system and in the majority of milking machines it is a rotary exhauster driven from either an electric motor or a stationary internal combustion engine, using pulleys and V-belts. This gives flexibility with speeds ranging from about 800-1500 rev/min and a corresponding ange of pump capacities to suit several sizes of milking machines.

A few vacuum pumps are direct shaft coupled; their speed is then motor speed and cannot be changed. When operating with a vacuum at the pump inlet suitable for milking cows (50 kPa), the pumping capacity should be sufficient for the total number of milking units connected. This will range from 70 litres of air per minute per unit for installations with 20 units to 85 per litres per minute for the smaller plants with 5 units.* It will vary according to individual manufacturers design of equipment and the additional ancillary components included. Oil lubricated high speed vacuum pumps operate at speeds of between 850- 1440 revs/min. For each 0.75 kW (1 h.p.) of rated electric motor requirement the extraction capacity is approximately 280 litres of air/min. The pump should be fitted with drive belt guards, an effective silencer and, for oil lubricated pumps, an oil separator. Also, a nonreturn valve should be fitted in the exhaust pipe to prevent reverse rotation of the vacuum pump when switching off at the end of milking. This is to avoid debris from the silencer/ exhaust pipe being sucked back into the pump. To prevent solid and liquid material being drawn into the pump from the milking installation, an interceptor/trap vessel with a capacity of not less than 15 litres is fitted in the main vacuum line adjacent to the pump, with provision for draining and cleaning. It should also have a float valve to shut off the vacuum in the event of it filling with liquid.

VACUUM REGULATOR

An automatic valve to maintain a steady vacuum inspite of varying air usage. A preset force (weight or spring) holds the valve closed until vacuum lifts it and allows air to enter. Valve should maintain vacuum to within the limits of ±kPa of the designed working vacuum.

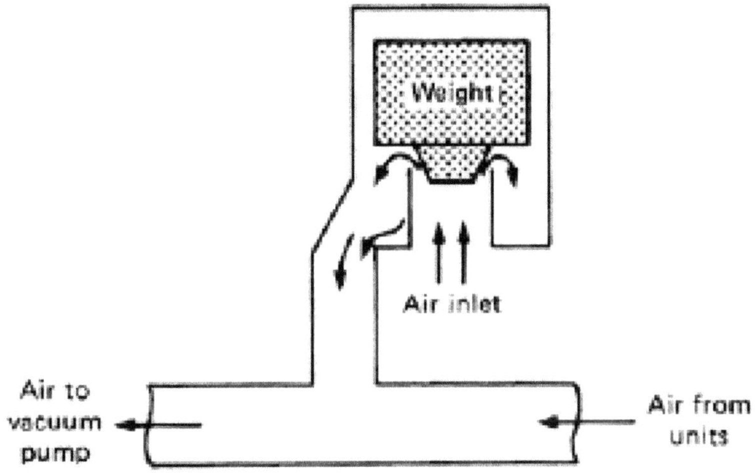

VACUUM GAUGE

Indicates abnormal levels and fluctuations in vacuum, eg. serious air leaks, dirty regulator and slipping vacuum pump drive belts.

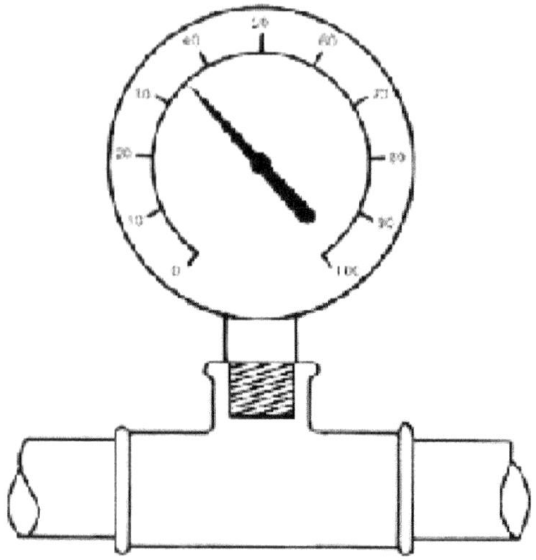

Vacuum regulator is an automatic valve fitted into the main vacuum pipeline near to the milking units that is designed to maintain the working vacuum level in the milking machine, despite the varying air usage during milking. It operates by admitting air into the pipeline when the vacuum increases above the predetermined level. The valve is held closed by a weight or spring until the vacuum in the system overcomes the closing force, allowing atmospheric pressure to open the valve. Weight operated regulators have a maximum capacity of approximately 550 to 700 litres/min. Spring operated regulators have capacities of up to 1400 litres/min. Servo or power operated regulators, which usually have remote sensing, utilize a small intermediate pilot valve to actuate the main air inlet valve. These regulators sense pressure changes in the vacuum pipeline at a position where more stable conditions might be expected to exist. They are readily adjustable with capacities varying from 1000 to 5000 litres of air/min.* A good regulator should be able to control the vacuum to within plus or minus 2 kPa of the working vacuum, be stable and not leak by more than 35 l/min, when nominally closed. All vacuum regulators have an air filter to filter the incoming air. The vacuum gaugeindicates the vacuum in the pipeline system. Its most important function is to indicate abnormal levels and fluctuations in vacuum, eg. serious air leaks, dirty regulator and slipping vacuum pump drive belts. Universally the Bourdon type gauge is used. It should not be less than 75 mm in diameter, graduated at intervals of 2 kPa and be adjustable. Most gauges have dual graduations, kilopascals (kPa) and millimetres of mercury (mmHg), and have lines on the dial indicating the working vacuum level.

PULSATORS

Pulsators are valves that cause the liners to open and close on the teat once each second (ie. pulsation) by connecting the pulsation chamber of the teatcup to vacuum or atmosphere.

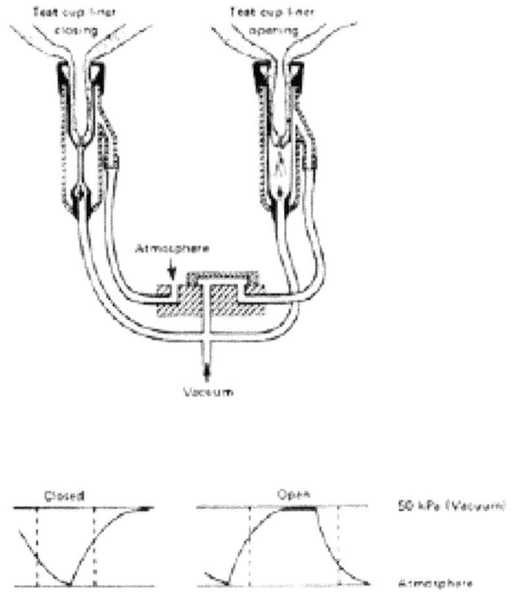

Pulsation chamber vacuum

The pulsator is a simple valve that alternately admits air and vacuum into the pulsation chamber formed between the rubber liner and he shell. This causes the liner to open and close during milking. The valve is activated either by a pneumatic (vacuum) or an electrical signal from a pulsator controller to give a frequency of 40-50 cycles per minute and a ratio (open to closed) of 1:1 to 2:1 (sometimes called 50:50 and 66:33). Relay pulsators are activated from a central pulsator controller, and self contained pulsators have built-in controllers. Most relay pulsators, electrical and pneumatic, pulsate the four liners of a cluster together (simultaneous pulsation). Self contained pulsators are better adapted to pulsating liners in pairs with two open and two closed (alternate pulsation). In the slide valve mechanism of self contained pulsators, the slide valve is arranged to produce a predetermined ratio. With a higher vacuum under the teat than in the pulsation chamber (sometimes due to milk flowing downhill) the liner will not open fully, slightly pinching the end of the teat, resulting in possible teatend damage. Conversely elevating the milk can reduce the vacuum under the teat, with the result that the liner does not fully collapse. This will influence the designed ratio of open to closed of the liner. These adverse conditions are minimised by ensuring that there is an air admission hole into the bowl of the claw to admit air within the range of 4-10 litres/min as recommended in milking machine standards. The pulsation rate of pneumatic pulsators is determined by a restriction in the air or oil damping system of the pulsator. With all pulsators, the size of the valve ports, length and bore of the pulse tubes determines the rate of extracting and admitting air. This determines whether a pulsator can operate more than one cluster and is a function of the basic design. There are however solid state electronic pulsators that control the rate and ratio which provide alternate pulsation and can operate two milking units. Air filters are fitted in the air intake to prevent dust getting into the alve mechanisms which should not be lubricated unless recommended in manufacturer's instructions.

Milking Cluster

Consists of four teatcup assemblies each having a rubber liner and connected to vacuum by rubber tubes and claw. The air admission hole to stabilise the vacuum must be kept clear.

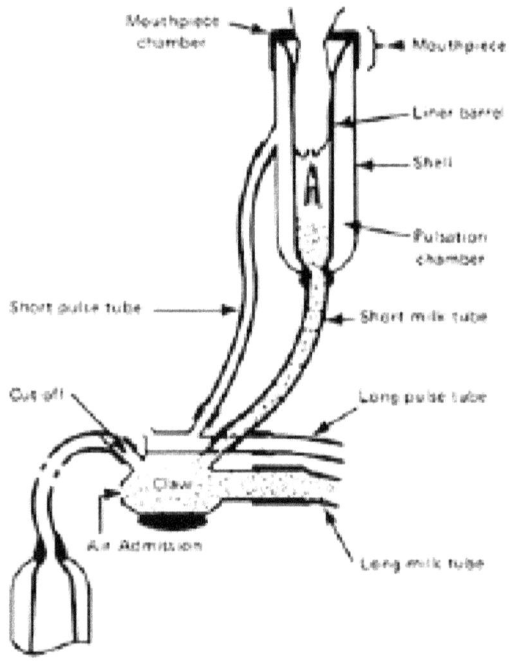

Cross-section of a teatcup cluster

Cluster

The cluster which attaches to the cow, consists of four teatcup assemblies (each having a shell, a rubber liner and a short milk and short pulse tube), a claw, a long milk tube and long pulse tube(s). Teatcup shells are normally made of stainless steel. Plastics or a combination of plastics and metal are also used. The liner is a flexible rubber sleeve having a mouthpiece, and when assembled in the shell under tension, forms an annular space (pulsation chamber) between the liner and shell. This pulsation chamber is connected to the pulsator through a nipple on the side of the shell via the claw. The teatcup assemblies are connected by short milk and short pulse tubes to the claw, which is connected to the milking and pulsation vacuum by a long milk tube and long pulse tube(s). To stabilise the vacuum in the teatcups during milking, the claw has a small air admission hole, about 0.8 mm in diameter, which admits approximately 7-8 litres of air/min into the bowl of the claw. This air helps to carry the milk away, preventing flooding and violent vacuum fluctuations. The claw is made of stainless steel or a combination of plastics and stainless steel, and usually weighs about 0.5 kg and the total all up weight of a milking cluster is about 2.5 kg. The weight of a milking cluster is important and the correct weight relates to the design of liners. Too little weight gives incomplete milking because of high levels of strippings, too much weight will result in milking units falling off during milking.

The bore of the rubber short milk tubes should not be less than 8 mm and the short pulse tubes not less than 5 mm, and the long milk tube should not be less than 12.5 mm. The effective claw bowl volume should not be less than 80 ml.

Milk And Air Separation

There are several methods of extracting milk from the vacuum system of pipeline milking machines.

The milk is collected in a receiver where the milk and air are separated ready to be extracted with:

a) a centrifugal releaser milk pump

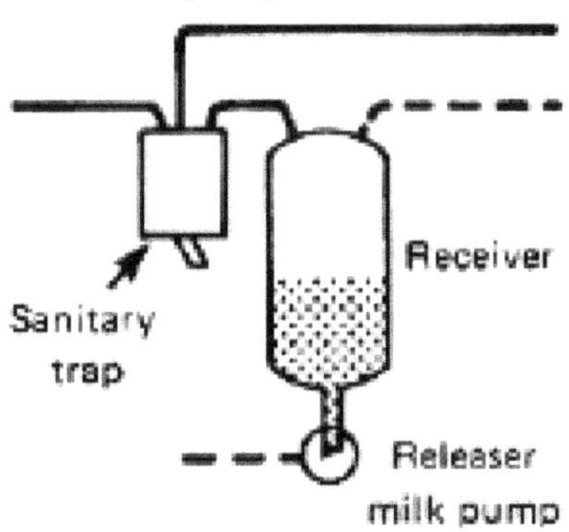

b) double chamber releasers with weight or float operated mechanical valves and the New Zealand pulsator operated 'spit chamber' releaser

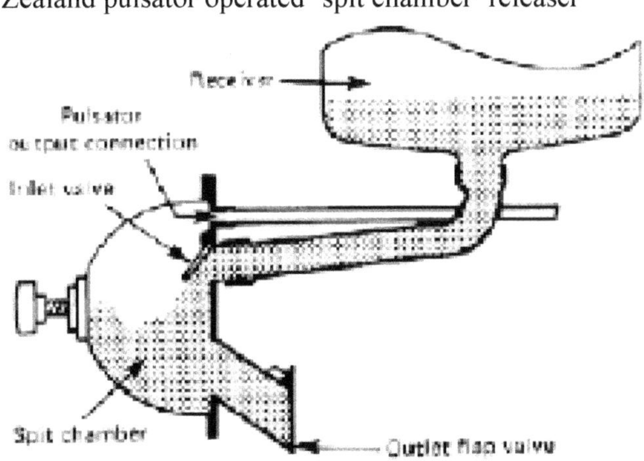

c. direct connection to a refrigerated farm vat (bulk tank) that is designed to operate under vacuum eliminating the need for a conventional receiver releaser.

In pipeline milking installations it is necessary to include a method of extracting milk from the vacuum system. A receiver vessel is fitted to act as a milk reservoir and air separator and from this the milk is pumped out. It is made of either glass or stainless steel and may have a capacity of 35 to 160 litres or more depending on the method of cooling and storing the milk. During milking the weight or level of milk in the receiver is used to start the milk pump (releaser). Releaser milk pumps for extracting milk out of the vacuum system are: centrifugal, with capacities of at least 4550 litres per hour, or diaphragm, with capacities of approximately 2000 litres/hr for a single ended pump. Other methods for extracting milk from a vacuum system are double chambered releasers with weight operated mechanical valves and the 'spit chamber' releaser. These do not require electricity and the latter utilises a pulsating vacuum to alternately open and close flap valves allowing the milk to drain out of the second chamber.

In addition to the interceptor for the vacuum pump, a sanitary trap, is fitted in the vacuum pipeline adjacent to the receiver. This is a glass vessel of not less than 3 litres capacity that separates the part of the milking machine through which milk passes from the air system, preventing movement of liquid from one to the other. If milk enters the sanitary trap it is an indication of a fault in the machine. Therefore, it should be mounted within sight of the milker and be fitted with a float ball to shut off the vacuum. The vacuum connection between the sanitary trap and the receiver should slope away from the receiver to the sanitary trap.

Recording And Sampling Milk

To measure the yield of each cow requires:-

a) the milk to be collected in a calibrated glass recorder jar.

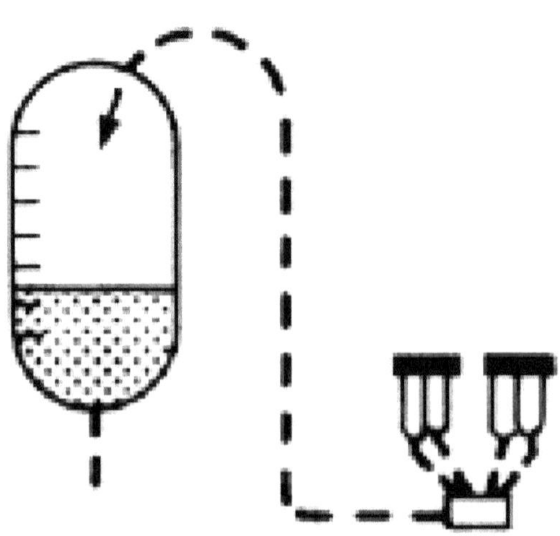

b) the milk to pass through a meter which can either take a constant proportion sample or to measure the whole yield in discrete batches and totalled.

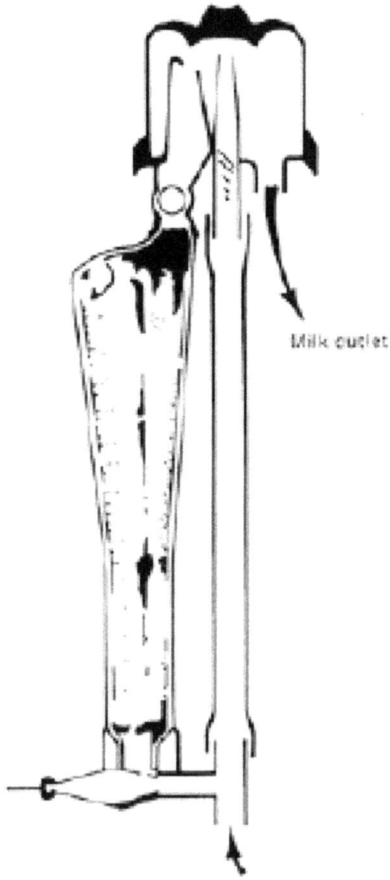

All methods have provision for taking a representative sample of the total yield for milk quality (fat, protein and lactose) analysis.

In pipeline recorder milking installations, milk flows directly from the cluster into a rigidly mounted calibrated glass recorder jar where the milk is intercepted and held to allow measurement of the total individual cows milk yield. This is done by using either the calibrations etched on the side of the glass jar or the weight of the contents for jars that are supported on electrical strain beams.

In direct to pipeline milking machines not fitted with recorder jars, milk yields are measured with milk meters. There are several designs of meters. Basically there are two distinct methods used. (1) Proportional flow; these collect a constant proportion (about 2.5%) of the milk flow into a calibrated flask the contents of which are recorded manually or electrically. (2) Batch; the total flow of milk is continuously measured in discrete batches and the aggregate of the batches recorded.

Proportional flow meters used for manual recording are relatively simple and inexpensive. Electrical recording of milk yields, depending on complexity can be costly and are usually linked with electronic management data recording systems.

All these methods of milk recording have the facility for collecting a milk sample for chemical analysis and are required to be within set limits of accuracy for yield and ability to collect a true representative sample of the milk. These limits are laid down in an International Standard for milk yield recording equipment by I.C.R.P.M.A.*

MISCELLANEOUS EQUIPMENT

2Milk flow detection

Devices are fitted to assist the milker to determine the end of milk flow. These can be:

- sight glasses or 'windows' in the claw or tubes
- glass recorder jars
- flow indicators which show visually when the flow rate falls below a definite level
- electronic equipment that measures change in flow rate

Mastitis detectors

- filters' in a transparent holder which indicate the abnormal milk

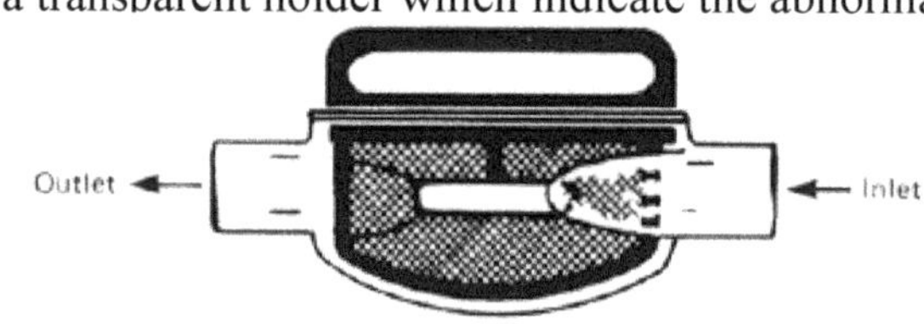

Teatcup liner 'shields'

These are fitted in the teatcup liners to reduce mastitis by preventing contaminated milk droplets impacting the teat ends when the vacuum fluctuates in the liner.

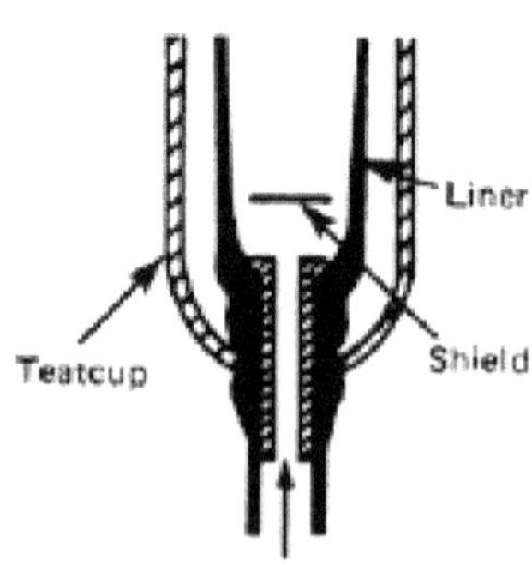

'One-way valves' (Non-return)

These are fitted in the claw, in the short milk tube or in the teatcup liner to reduce mastitis by preventing reverse-flow of milk

Milk Flow Detection and Cluster Removal

With all milking units, except those using transparent recorder jars, there is some provision made for observing milk flow. This can be a metal claw fitted with "windows", or a claw made in a transparent material or, a short length of clear tubing fitted in the long milk tube. Observing flow in these ways can be misleading and there are now milk flow indicator devices that work on the basis of milk flowing into a small container with a submerged metering orifice for maintaining a milk level. Variations in this level can be observed or detected electrically, using a magnetic float switch or electric probes.

In modern large milking installations, milk flow detectors are used to sense the end of milk flow (less than 250 g/min) and activate automatic cluster removal (A.C.R.) equipment. This equipment prevents overmilking and enables the operator to use more units by reducing the workload. It eliminates waiting to observe when an animal has finished milking and manually detaching the cluster. Electrically operated milk yield recording equipment eg. glass recorder jars on strain beams and milk meters can be used as flow detectors and linked to A.C.R. equipment. In this way the milking cluster is removed when flow rate has fallen below a predetermined level. A useful device that can be fitted in the long milk tube is a small in line filter used as a mastitis detector. It retains milk clots enabling clinical mastitis to be detected and also acts as a coarse filter for removing extraneous material which is an advantage when using flow detectors or milk meters that have small metering orifices etc. The filter screen is easily removed for rinsing off the clots after each cow and the detector is cleaned in situ by the normal cleaning system of the milking machine.

To reduce or eliminate the effect of vacuum fluctuations that occur during milking "shields" can be fitted in the bottom of the liners near the outlet preventing contaminated milk returning up the short milk tube from impinging on the teat end. The shield consists of a disc of stainless steel or plastics of a diameter that leaves a space etween the edge of the disc and the liner wall not less than the internal diameter of the milk outlet. A more positive method to prevent the reverse flow in a conventional cluster is to fit non-return valves in the claw or short milk tubes together with a small air admission hole between the liner and the valve to maintain normal milk flow. A system of milking with non-return valves but without the air admission holes has been investigated and preliminary results show the development could have advantages.

Water Heaters

The quantity of hot water required depends on the size of milking machine and the cleaning methods used. It is essential to have an adequate supply. The methods of heating can be, a solid fuel or oil fired boiler, an open top un-insulated electrical wash boiler or a self contained insulated electric water heater. Complete with time switch, thermostat and automatic water filling.

The electrical heating element should be fitted with a clear space beneath it for the hard water scale that will collect.

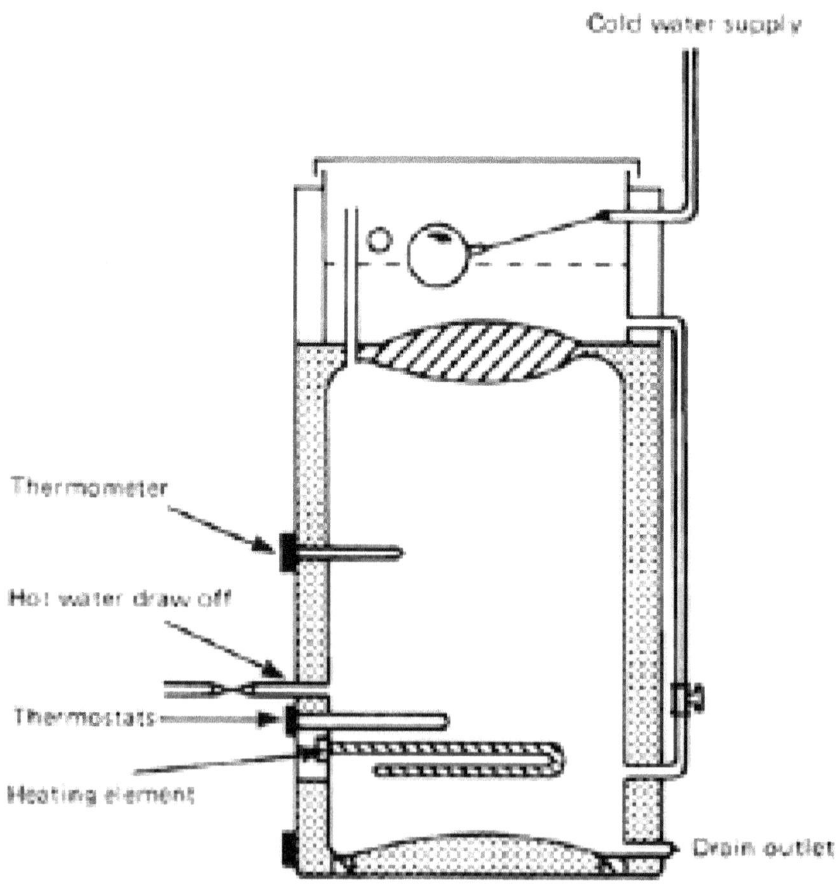

Water Heating

For nearly all milking machine cleaning methods an adequate supply of hot water is necessary. Water heaters vary from simple free standing insulated or uninsulated tanks filled with a hose and manual control to insulated tanks automatically filled and heated at predetermined times.

The water capacity required is dependant on the washing system used (hand or in-place cleaning) and the number of units to be washed. For in-place cleaning the amount of water required for each milking unit will be 12 literes of water at 85°C for recirculation cleaning and 18 litres at 96°C for a single pass "boiling water" cleaning system. Additional hot/warm water may be required for udder washing and calf feeding. Whatever cleaning methods are used the water heater should be capable of producing near boiling water for the purpose of giving the installation a periodic heat treatment. The temperature of the wash solution in closed pipe work is affected by the vapour pressure of the liquid. Water at normal atmospheric pressure at sea level boils at 100°C (212°F) and at a vacuum of 50 kPa (14.8 inHg) it boils at 81.7°C (179°F). Thus when cleaning pipeline milking machines with water solutions under vacuum, the maximum temperature which can be obtained ill depend on the operating vacuum level in the installation, ie., the higher the vacuum level the lower will be the temperature that can be obtained. For similar reasons, at an altitude of a 1,000 metres where atmospheric pressure is reduced to about 89 kPa (12.9 lb/in2), water in an open vessel will boil at the lower temperature of 96°C (205°F) instead of 100°C (212°F) at sea level.

Although any fuel can be used for water heating the most commonly used is electricity for cleanliness and convenience. In areas with hard water supplies, a hard scale will build up in the tank and on the heating element. The electrical heating element should be fitted with a clear space beneath it (approx. 10 cm) and be of a type that will flex during heating and cooling, breaking off the hard water deposits that will fall into the space below. Provision should be made for easy removal of this accumulated scale.

For economy any water heater should be well insulated, fitted with a thermostat and preferably a time switch. In soft water areas tanks that are manufactured in galvanized sheet metal are not suitable and copper should be used.

Maintenance

For effective milking giving good yields of high quality milk, fast milk removal and to minimise mastitis, good milking machine maintenance is required. Maintenance routines should be provided by the manufacturers of the equipment.

These will include

- simple checks to be made immediately before each milking
- checks at weekly and monthly intervals
- an annual check of the complete plant by a trained technician.

FAULTS FOUND MUST BE CORRECTED

Otherwise the checks are of no value.

Maintenance

The basic layout and operation of milking machines is normally straight-forward and similar for all standard milking machines. However, it is important that a correct maintenance routine is followed. Faulty milking machines can result in poor milk let down, slow milking, milk of high bacterial count and mastitis. Maintenance is a responsibility of the person regularly using the equipment with simple checks made at each milking and more detailed ones at weekly or less frequent intervals.

The maintenance instructions described deal with the essential routines and do not include checks on the many optional items not fitted in large modern parlours, (eg. automatic cluster removers, milk meters). These vary considerably and the maintenance procedure should be obtained from the supplier. Because the overall performance can be assessed only with special measuring equipment a fullscale test should be carried out by the manufacturer, installer or extension service at least once per year.

The following routines should be read in conjunction with maintenance procedures provided by the suppliers of the milking equipment.

Maintenance Checks

At each milking

check pipelines and interceptor are free from milk or water, if found, drain and flush with chlorinated water (100 ppm)

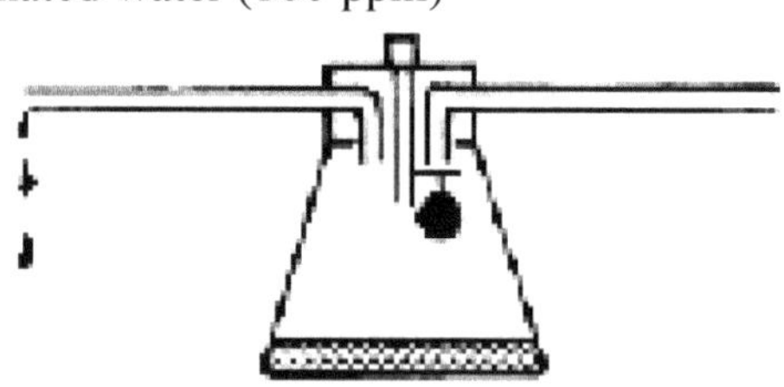

- check for water behind teatcup liners if found drain
check oil level in pump, if necessary top-up to correct level

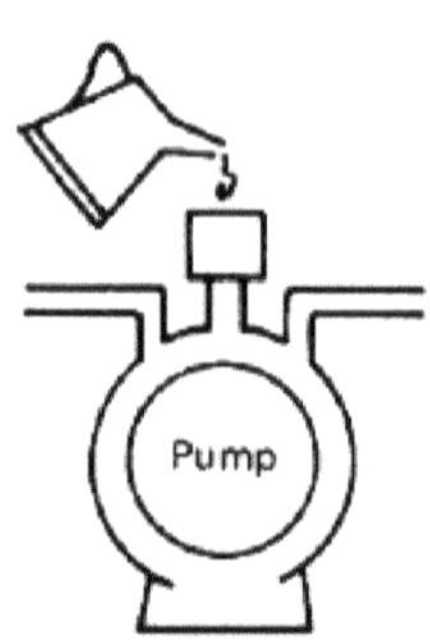

- check vacuum level and rate of recovery, if level too high check vacuum regulator, if recovery is slower than usual to reach working level, look for leaks into machine

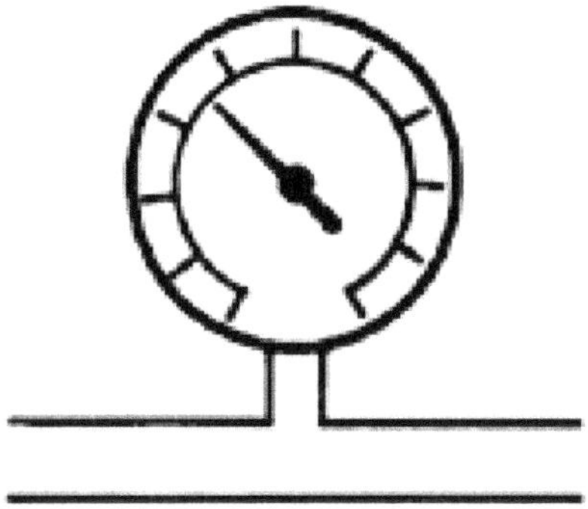

- during milking listen to ensure that regulator is continuously letting air into the vacuum system

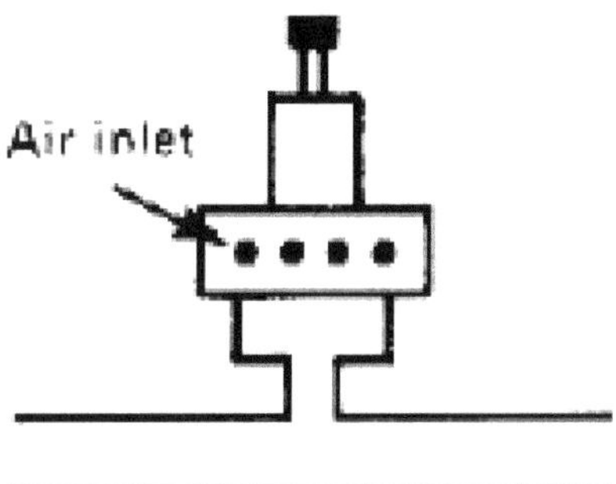

- check air admission hole in claw is clear

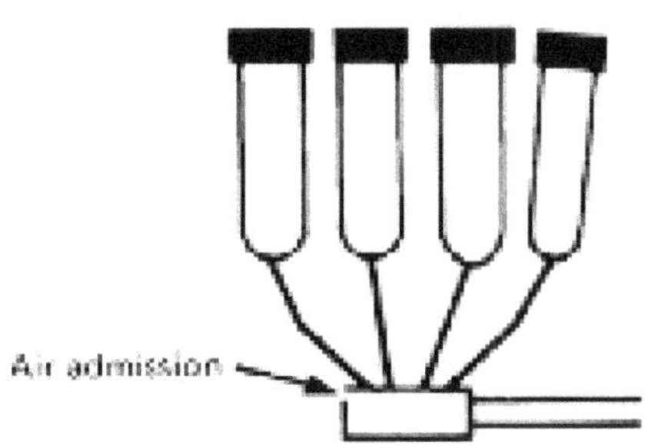

At Each Milking

Confirm that wash water is not retained in milking machine pipelines, particularly the milk lines. This water is likely to have a high bacterial count and reduce the quality of the milk. If water is found drain and flush the milk lines with hypochlorite solution (100 ppm) before milking. In most countries it is an offence to add water to milk and residual water in a milking machine may be sufficient to fail the freezing point of milk test. With some machines and washing systems water can enter the pulsation chambers of the teatcup, damage pulsators, affect the pulsation of the liner and gain entry into the main vacuum systems giving high levels of bacterial contamination. Check the clusters for split liners.

Before starting the vacuum pump confirm that the interceptor is empty either by opening the drain valve or detach the trap for inspection.

Confirm that the oil level in the vacuum pump oiler is at a satisfactory level, ie., not less than half the capacity of the oil reservoir.

Observe the vacuum gauge, the working level should be reached within five seconds, or consistant with the time stated by the manufacturer when the equipment was installed. If there is a delay bove that expected, check for air leaks at open stall or drain cocks, a misplaced receiver or bucket lids and where a centrifugal milk pump is used, for failure of the non-return flap valve to close. If all these items are in order check that the vacuum pump drive belts are not slipping. The working vacuum should be constant and the reading on the vacuum gauge should be the same before and after putting the milking clusters on the cows. Faults found should be corrected at the earliest opportunity.

- listen to ensure normal sound from pulsators. Check suspect action with thumbs in teatcup liners

After each milking

- allow pump to run for 10 minutes after end of milking

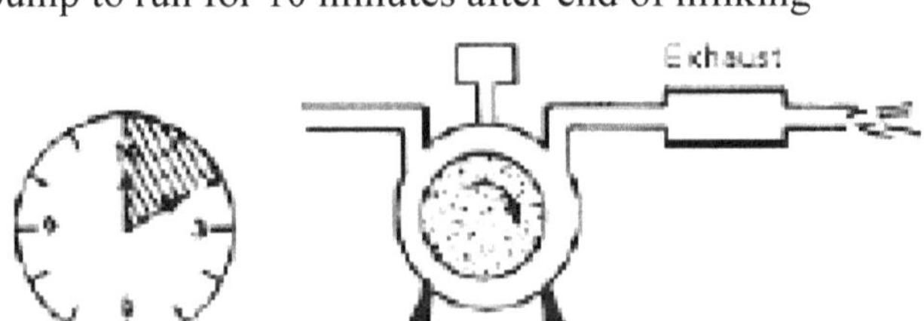

- check for milk in interceptor. If found find cause and wash out pipeline and interceptor with a warm detergent solution.

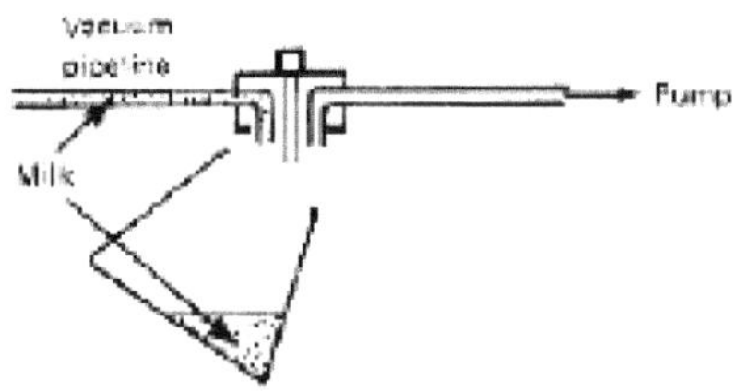

When all units are in operation and the working vacuum is correct, listen to the vacuum regulator. If there is no sound of air entering the regulator, a serious shortage of pump capacity is indicated. If on starting the vacuum pump the vacuum level rises first to a high level then drops to the working vacuum level the regulator is sticking and requires cleaning.

Check operation of the pulsators. The sound from the pulsators should be regular. When noise is abnormal look for damaged/split rubber pulse tubes particularly at the nipples on the claw, or faulty pulsators. If there is any abnormality check pulsation of liners with thumbs (see Weekly Check).

Examine the milking cluster for damaged rubberware, particularly the short milk and air tubes at the claw nipples.

Condensation or moisture from milking or plant cleaning can collect in the vacuum pump. This rusts the pump body producing, a condition which reduces performance and ultimately would require pump eplacement. This can be avoided by allowing the pump to run for approximately 10 minutes after milking to exhaust the moisture and dry out.

After switching off the vacuum pump examine the interceptor. If milk is found in the interceptor, this may indicate a split liner or malfunctioning of the milk pump. Find the cause, wash out vacuum line and interceptor.

MAINTENANCE CHECKS

Weekly

- check pulsator action of all clusters with thumbs in teatcup liners

a) for pulsation ratio of squeeze to release for number of pulsations per minute

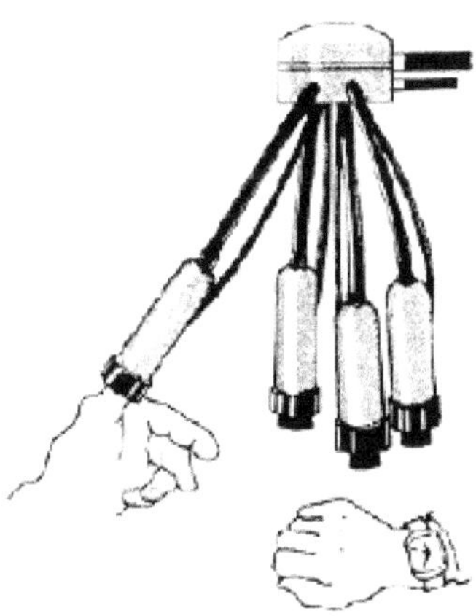

Inspect all rubber tubes for kinking, flattening or damage, in particular short pulse tubes and milk tubes for holes, replace if necessary.

Weekly

To test the pulsation and liner wall movement use the thumbs to sense the movement of the teatcup liner walls. This is done by inserting the thumbs into diagonal pairs of teatcups in turn with the others hanging in the cut-off position; the controls being set for milking. To make certain that the pair of teatcups not being checked are sealed they can be closed with rubber plugs (bungs). With experience abnormal action of the liner is readily detected. If the liners open and close completely and the action of the machine is not painful it will normally be satisfactory for milking. Failure to collapse indicates that the air inlet to the pulsator is blocked or that the pulsator valve is not working properly and the pulsator must be replaced. If this is discovered the operator will find that the cows arc uncomfortable and restless and kick clusters off. The ratio of the liner open to closed cannot be confirmed by this test but it should be possible to detect that the collapse or closed phase is shorter than the open, milking phase. If the pulsation appears to be unsatisfactory check for damaged short and long air tubes.

When checking the liner wall movement confirm that the speed of pulsation is correct. Slow pulsation will give longer milking times. In self-contained individual pnuematic pulsators this slow action is usually due to a low vacuum level or a dirty slide valve. Some designs f pulsators have a speed adjustment facility, this should be reset if incorrect. Most designs of pulsators are designed to operate in the range of 50-60 pulsations per minute. The pulsation speed can also be checked during milking by a slight pressure on a short pulse tube with thumb and fore finger.

- a slack vacuum pump drive belt will result in wear and loss of vacuum. Check tension by deflecting belt, movement should not exceed 15 mm (0.6 in). Also check oil level in vacuum pump oil reservoir and refill.

- inspect air filters on pulsators and vacuum regulators. Clean off light dust deposits by brushing or blowing. Heavily contaminated foam plastic filters should either be renewed or washed in warm detergent solution.

Vacuum Pump Drive Belt And Oil Level Of Pump

Check the Vee drive belt tension on vacuum pumps. Deflect the belt downwards at a point midway between the pulleys. The movement or deflection possible from the rest position should not exceed 15 mm (0.6 in). A slack drive belt will slip, reducing the vacuum pump performance. If serious belt slip is occuring there will be visible particles worn from the belts on the floor. If adjustment is needed care should be taken to keep the pulleys in alignment. Check the oil level in oil reservoir and if necessary refill.

Filters

Air filters on pulsators and vacuum regulator must be kept clean. The frequency of cleaning depends on whether they are operating in a clean or dusty environment. Most modern components are fitted with foam plastics air filters or screens. Brushing or blowing will remove light dust deposits. Heavily contaminated foam plastics filters should either be renewed or washed in a warm detergent solution and thoroughly dried before re-assembling into the pulsator.

Monthly

-dismantle and clean weight operated vacuum regulators. Using a soft brush and a solvent, methylated spirit. Taking care not to damage valve surfaces.

Remote sensing regulators

- follow manufacturers instructions.
- wash out vacuum line and interceptor, using hot detergent solution and drain. At the same time check interceptor for leaks, from rubber seals or small holes in base.

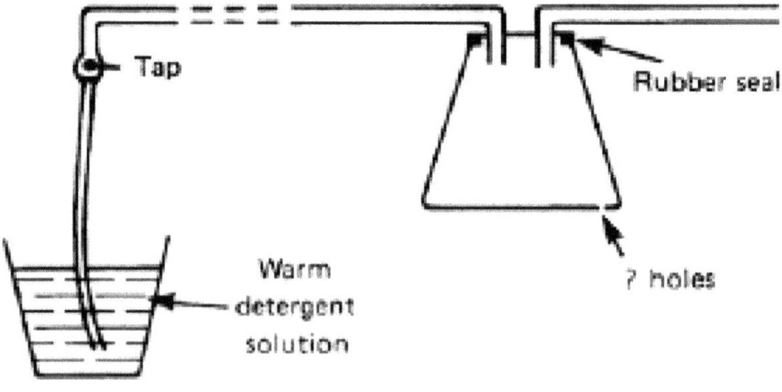

Monthly

Dismantle weight operated vacuum regulators, clean valve face and valve seat with a solvent (methylated spirit). Care must be taken to avoid amage to valve surfaces. Remove difficult deposits with metal polish. Clean regulator filter before the regulator is reassembled and ensure that the weight is evenly balanced.

Servo or power operated remote sensing designs of regulators should be serviced according to the manufacturers instructions as correct reassembly is not always easy. They will require the conical valve face to be cleaned and rubber diaphragm(s), if applicable, inspected. DO NOT oil valve or valve guides.

Check that damping rings or rubbers etc are not damaged.

Particular care must be taken when reassembling Servo Regulators. Observe closely during dismantling how rubber diaphragms, pilot valves etc, are fitted, so that they can be assembled correctly.

The vacuum line and vacuum pump interceptor trap should be washed through with hot detergent-hypochlorite solution. Make up a wash solution of detergent in hot water (10-12 litres at 60-70°C) With the addition of hypochlorite (250 ppm). In the case of branched lines, this quantity should be drawn in along each branch. Provision should be made to enable this to be carried out easily by providing at the end of each pipeline a cock or valve with a nipple suitable for attaching a rubber tube to suck the liquid from a bucket into the pipeline. When sucking the solution into the pipeline, allow air to be drawn in at the same time by withdrawing the tube from the solution at frequent intervals. The air liquid mixture will achieve a scrubbing action in the pipeline and be more effective than a solid stream of liquid. Most interceptor vessels are about 18-20 litres capacity and must be emptied after each 12 litres of cleaning solution has been drawn into the plant. After washing the lines the drain cocks must be left open to allow the vacuum system to drain. Before re-installing the interceptor, check the rubber eals and observe if there are any small holes in the base of the interceptor where air leaks can occur. These can cause a carry over of liquid from the interceptor to the detriment of the vacuum pump.

• carry out manufacturers maintenance instructions for PULSATORS Pneumatic self contained pulsators, if badly worn, can usually be made good by replacing worn components, ie. rubber diaphragms, valve slide and spring. Exchange reconditioned pulsators are available.

• check for air leaks at MILK PUMP rubber non-return valve or shaft seals. With water in the receiver jar and under vacuum look for air bubbles in jar. If present dismantle and find cause.

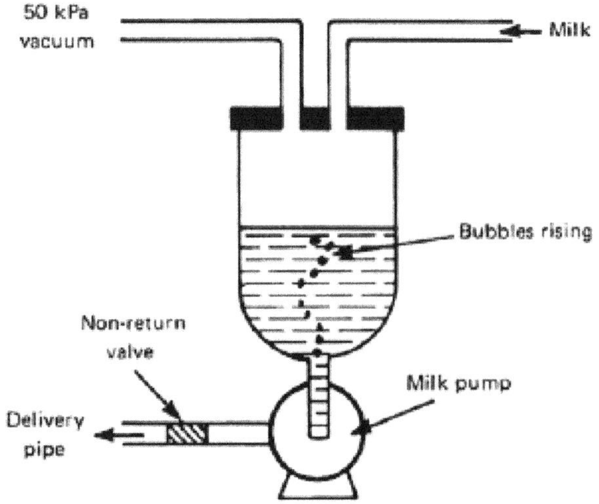

Pulsator maintenance varies depending on type, (electronic, electro-pneumatic and pneumatic) and whether relay or self contained. Manufacturers instructions should be followed. Filters, air inlet ports and in the case of pneumatic pulsators, the valve and valve slides require cleaning.

A pulsator performs about 6 million operations a year and is therefore subject to considerable wear. A badly worn pulsator can usually be made to operate satisfactorily by replacing the worn components with new parts supplied by the manufacturer e.g. rubber diaphragms, valve slides and springs. It will normally be necessary after two or three years to replace pneumatic pulsators and pneumatic relays. Manufacturers may have an exchange service providing reconditioned pulsators.

Milk Pumps

The performance of a centrifugal milk pump can be seriously affected, even made inoperative, by air leaks at the rubber seal on the drive shaft or the non-return flap valve fitted in the pump outlet or delivery pipe. Where glass receivers are in use, test for leaks by putting approximately 4 litres of water in the receiver with the milk pump not working. With the receiver vessel under vacuum look for air bubbles rising to the surface from the receiver outlet. If a continuous stream of bubbles is seen dismantle the non-return valve and examine rubber flap for wear and distortion, renew if necessary. If leaks continue replace pump or dismantle and renew shaft seal, inspecting shaft for wear. It will be necessary to replace pump if there are signs of annular grooves on shaft.

Routine replacements

- Renew milking machine liners every six months, or 2,000 cow milkings.
- Replace long milk tubes every year.
- Replace long pulse tubes, rubber elbows and connectors every two years or when damaged.

Yearly

A full test of the milking machine must be carried out once a year by an experienced technician as laid down in the International Standards; ISO 5707 “Milking Machine Installations - Construction and Performance” and ISO 6690 “Milking Machine Installations - Mechanical Tests”.

10
Dairy Heifer Management

Raising dairy replacement animals is a challenging job in that heifer raisers are responsible for the milking herd's future-the next generation of milk cows. The cost of raising dairy calves and replacement heifers is the second highest expenditure on the farm, behind feed costs. It is a major component in making a dairy operation profitable, but yet sometimes is neglected as a profit center for the farm since no direct income is generated by the replacement herd.

MANAGING GRAZING HEIFERS

When yearling heifers graze with older cows, they become exposed to and hence developing immunity against other diseases carried by these animals. This immunity is then transferred to their offspring via their colostrum. If reared in complete isolation, heifers risk becoming infected after they calve and first come into contact with older cows. This is a time when they should be in peak health to produce milk, get back in calf and adapt to their radical change in management. Prior to first calving, heifers are relatively unstressed as they are only occasionally yarded and spend most of their time at pasture. On entering the milking herd, they experience more competition for feed and social position in the herd. Careful monitoring is required to ensure that they still achieve target live weights, while competing with older cows for limited pasture over winter.

Young stock are inquisitive. They should be kept away from steep-sided drains, cliffs and other dangerous locations. They may also try to eat wire or baling twine, which can injure them. Young stock should be handled frequently. After calving, it will be easier for them to adjust to milkings and find their place in the herd's social structure if they are used to human contact. Hand feeding them for their last few weeks of pregnancy provides extra feed to build up body condition and also adapts them to handling. For example, they could be fed while run quietly through the milking shed.

Animal Health

It is important to plan and undertake routine control for both internal and external parasites, and vaccinate against diseases, using regional recommendations. These include drenching for round worms and liver fluke and vaccination against Clostridia and Leptospirosis.

Vaccination for Clostridia is vital and cheap. Calves should be vaccinated at weaning, with boosters given six weeks later, at one year of age and again at drying off. The last vaccination ensures some immunity is passed through their colostrum. Clostridial diseases will become more common as young stock are fed more concentrates and achieve better body condition.

Worm burdens can build up on pastures where heifers are set stocked for lengthy periods. Routine faecal egg counts should be undertaken to check for sub-clinical parasite infestations. The common symptoms of clinical parasite infestations include ill-thrift, roughness of coat, scouring and coughing.

Cattle under the age of 12 months are susceptible to Johnes disease and is important that such heifers are excluded from pastures previously grazed by older stock. Johne's disease bacteria are shed in the faeces of cattle over two years of age and can survive up to 12 months in cool, moist conditions. This is reduced in warm, dry conditions or when the bacteria are directly exposed to sunlight.

Heifers should be regularly checked for pink eye and bloat. Mineral levels in grazed pastures are usually sufficient to achieve target growth rates. However, additional minerals may be required if high levels of supplements are fed, particularly cereal grains.

Heifers can develop their first permanent incisor teeth any time between 22 and 28 months of age, although the majority of heifers do not show teeth eruption until after calving. However, well-grown heifers can show teeth eruption prior to calving. When the temporary "milk" teeth fall out, with the newly erupting permanent teeth only just showing through in the gums, their efficiency of grazing short pasture may be impaired. It could also reduce their ability to compete for available feeds during their second winter. If individual heifers are not growing or do not compete well at pasture, their mouths should be inspected for the condition of their gums and teeth. Some may have loose teeth, which can be pulled out, or ulcers or infections, which can be treated. Preferential feeding of these heifers is advisable.

Reproductive Management of Heifers

For farmers to achieve the greatest genetic gain in their herd, they should mate their maiden heifers by artificial insemination (AI) using semen from proven bulls of known Australian Breeding Values (ABV's) for milk, fat and protein and a rating for ease of calving. The advantages arise from:

- production benefits from their progeny
- reducing the generation interval
- increasing the number of calves of higher genetic merit available as herd replacements
- reducing the number of bulls required on the farm
- a shorter mating program, resulting in a shorter calving period and less time spent on calf rearing

The traditional management system of naturally mating with terminal sires leads to a 5 year interval between the birth of replacement heifers to the entry of their progeny into the milking herd. If heifers are mated to AI at 15 months of age, their first calves will enter the herd when they are 4 years of age.

A herd of 100 cows requires 25 to 30 heifer replacement calves each year. If beef bulls are used on the maiden heifers, these replacement calves must be bred from nearly all the early cycling cows, allowing little chance for selection of better cows for future replacements. Using AI over the heifers as well provides greater scope for selection and genetic improvement of the herd.

Ease of calving depends on the condition of the dam and calf factors such as gestation length, calf weight and presentation at birth. A bull rated highly for calving ease can still produce calves that have the potential to grow into large framed cows, even though they should have lower birth weights. When using AI, it is important to select easy calving sires. These should have "calving ease" values of 2 to 3%, that is, only 2 to 3% of the calves born to that sire in mature cows will present calving difficulties.

Heifers should be gaining weight for at least 60 to 80 days before mating, have a condition score of 5 or more, and be seen to be cycling. Heifer heats are much less obvious and of shorter duration than those in older animals. An 8 hour heat period is not uncommon, so successful heat detection in an AI program requires the input of time from the farmer. Tail paint, Kamar pads and teaser bulls with chin ball harnesses are useful aids for heat detection.

Friesians show puberty at about 240 kg live weight, but this can occur at any age between 8 to 20 months. Target live weights for Friesian heifers at mating are 300 to 360 kg, but they must be at least 280 kg to ensure reasonable success. Corresponding live weights for Jersey heifers are 240 to 275 kg targets with a minimum of 210 kg.

Heifers in poor condition, and losing or only maintaining weight will have reduced fertility. Not only will smaller heifers have poorer conception rates, they can also have a higher incidence of calving difficulties. Live weight at mating is a better predictor of calving difficulties than live weight or body condition at calving.

Oestrus synchronisation in heifers

A synchronisation program can reduce the AI period to one or two days, with obvious management benefits. Such a program would also "tighten up" the calving pattern, reducing the days required for frequent observation of calving heifers.

In seasonal calving herds, a heifer synchronisation program advances the average calving date up to 10 days. This is partially offset in AI programs by reduced submission rates. If the heifers are mated say, 10 days before the cows, then their earlier calving date will give them 10 extra days prior to the start of the next mating. This is important for first calf heifers as they take longer to recover from the stresses of calving and commence cycling.

Use of AI with maiden heifers is most practical following a program of oestrus synchronisation. Results are better under ideal conditions, such as during fine weather, with access to good yards and handling facilities and when heifers are in good body condition.

Three main systems are currently used:

- CIDR or controlled internal drug release. These devices release progestagen (synthetic progesterone) and are often used with oestradiol capsules at insertion or with a prostaglandin injection on removal.
- Syncro-mate B pellets as ear implants. These also release progestagen and are often used with oestradiol injections at the time of implant insertion.
- Prostaglandin injections (Lutalyse or Estrumate). These systems usually require two injections, 10 to 15 days apart.

None of these system can guarantee tight synchrony, where the ideal is a single AI at a fixed time after treatment. Experience has shown that best results are usually achieved with insemination on heat detection 2 to 5 days after treatment. The programs require multiple yardings of heifers, so they must be quiet and used to being handled. Hygiene is also critical.

Oestradiol and prostaglandins are schedule 4 (S4) drugs and so can only be supplied by veterinarians, whose responsibility it is to prescribe the program and drugs used.

Some treated heifers may produce an unpleasant discharge at CIDR removal, although this is unlikely to adversely affect fertility. A few heifers may also suffer damage from CIDR's due to pressure on the vaginal wall.

The biggest single cause of failure in synchronisation programs with maiden heifers is low mating live weights. Such programs cannot tighten extended calving patterns in herds, since they have no effect on non-cycling heifers or cows. If natural mating is preferred, the synchronisation program can be staggered over several days. Some farmers may prefer natural mating of heifers, especially if they do not have the time to spend observing heats and handling them. Whatever the system, heifers should be mated to "easy calving" bulls. Heifers which have difficult calvings can suffer significant growth checks and are more likely to conceive late or not at all during their first lactation.

Despite attention to achieving higher live weights at mating and the publication of calving ease values for AI bulls, many farmers still find that Friesian heifers mated to Friesians leads to too many difficult calvings, compared to AI or natural mating with Jerseys. They may still prefer to use Jerseys for mating their Friesian heifers, either for ease of calving and/or to produce herd replacements. This is a practical option as herd recording data show that Jersey x Friesian cows can be very competitive in milk production, and may even have superior feed efficiency, after taking into account their lower live weights.

To ensure good results with natural mating, bulls should be physically examined for reproductive abnormalities and have their feet trimmed at least 6 weeks before the mating season. They need to run at the rate of one bull to every 30 heifers expected to cycle, plus one extra bull in the breeding herd. The numbers can be reduced as more heifers become pregnant. Any heifer not in calf after a three month mating period should be culled.

Maiden heifers often produce low quality colostrum. If so, their calves should receive additional or alternative colostrum from older cows. If the heifers are calved down before the older cows, this can be supplied from a frozen colostrum bank, from neighbours who calve their cows down at the same time or even by inseminating selected cows to calve down with the heifer herd.

Rearing Dairy Replacement Heifers

Rearing calves and heifers as replacement milkers, is a serious business and the replacement stock are one aspect of dairying no dairy farmer can afford to overlook. In brief, successful calf- and heifer-rearing depends on animals attaining target weights and heights for age, using a cost-effective system. While it is true that the milking herd generates the income, calves and heifers are just as valuable to the dairy's future as the current milking herd. Remember, today's calves will enter the milking herd in two years time. Raising calves takes special skills and knowledge. While the responsibilities of calf care are great, so are the rewards. Assuming that anyone can successfully raise calves is as costly as assuming the end is closed on the bulk tank. Poor calf management allows future profits to slip away. Following a few basic, but very important steps, will help ensure that every calf matures into a profit-generating cow.

To successfully raise calves, emphasis should be placed on the following:

- Minimise stress by good management - only trained, reliable staff who enjoy and take an interest in the calves should do calf chores.
- When more than one person feeds calves, it is vital that all communications are clear and that the same procedures are followed daily.
- Plan ahead and prepare thoroughly before calves arrive at the calf barn/shed.
- Use shelter effectively to protect calves from the weather, while minimising disease problems.
- Select calves that will grow well (bought-in calves).
- Use the right feed balance at the right time for maximum benefit.
- Start preparing the calves for the dairy-feeding system early, i.e. TEACH THEM TO EAT.
- Keep live-weight gain as high as economically possible up to and past weaning.
- Disease management – appropriate prevention, detection and treatment.

Calving Checklist

Once a cow has calved :

- Check that the calf is breathing. If the calf does not begin breathing, stimulate the inside of the nostril by tickling with a piece of clean straw, sneezing will force the calf to take a breath! Massaging the calf's chest also stimulates breathing.
- Allow the cow to lick the calf, thereby stimulating the new-born's circulatory system and drying the calf's coat.

- Milk the cow for colostrum.
- Dip the calf's navel cord in a 7% tincture of iodine solution. Completely cover the cord and the hair on the belly. Only use an iodine preparation specifically for dipping navels. Teat dip and preparations for humans are NOT strong enough to disinfect a calf's navel effectively.

Colostrum

The importance of colostrum cannot be over-emphasised. Colostrum is the thick, creamy, yellow secretion collected from the cow's udder after calving. By definition, only milk from the FIRST milking after calving can be referred to as COLOSTRUM. Milk secretions from the second to the eighth milking (5 days post-partum) are called transitional milk, as the composition gradually becomes similar to that of milk.

Table 1 : Composition of Colostrum

	DRY MATTER %	FAT %	TOTAL PROTEIN %	CASEIN %	ANTIBODIES %	LACTOSE %	MINERALS %
1st colostrum	21.1	6.7	14.0	5.72	7.33	2.9	1.1
After 24 hours	19.3	5.5	9.2	4.78	4.52	3.5	1.0
After 48 hours	15.5	5.4	5.0	3.38	1.62	4.2	0.8
After 72 hours	14.5	5.0	4.4	3.27	1.13	4.4	0.8
20th day : milk	12.9	4.1	3.4	2.66	0.74	4.9	0.7

Colostrum has a high nutritive value, plus provides the calf with ANTIBODIES needed to protect the newborn calf from infections that may cause diarrhoea and/ or death. Antibodies, or immunoglobulins are proteins normally found in the bloodstream and are vital components of the immune system. They help identify and destroy bacteria and other foreign particles (antigens) that have invaded the body. Antibodies are not present in the newborn calf's bloodstream because they cannot cross the placenta during pregnancy. The antibodies are transferred to the calf via the colostrum, where they are then absorbed through the intestine.

Quality of the colostrum is determined by three factors:-

Length of the dry period : A short dry period (less than 4 weeks), premature calving, milking before calving and milk leakage before calving will lower the concentration of antibodies in the colostrum.

Age of the cow : Generally, the concentration of antibodies is higher in older cows than in first-calf heifers. Older cows also produce colostrum with a more diverse antibody population than younger cows because of the length of time they have had to build immunity to diseases existing in the herd.

Breed of dairy cow : Holsteins have a lower antibody concentration in their colostrum than breeds such as Ayrshire, Brown Swiss, Guernsey and Jersey.

Do not rely on the cow to nurse for colostrum, as the calf may suckle but not ingest colostrum, especially in cows with pendulous udders or excessively large teats. The more high quality colostrum one can get into the calf, when it can be best absorbed, the better…don't worry about overfeeding colostrum at the first feed. High mortality rates are linked to too little quality colostrum intake, not too much. Most calves that die, do so because they do not have the ability to resist a virus challenge and fight common infectious diseases.

The amount of colostrum fed and time of feeding in relation to birth considerably influences calf survival. Immediately after birth, ABSORPTION of antibodies is highest, with a rapid decline in the efficiency of absorption of antibodies as soon as 6 hours after birth. Thereafter, DIGESTION of antibodies increases and intestinal cells rapidly become impermeable to antibodies. Approximately 24 hours after birth, calves lose the ability to absorb intact antibodies. It is still beneficial to feed excess colostrum to calves as the antibodies (gamma globulins), although they will not be absorbed, still have a valuable disease-controlling potential by neutralising pathogens locally in the gastric mucosa (stomach lining).

The First Feeding

Collect colostrum from the mother soon after calving, after the udder and equipment has been properly cleaned and sanitised. For the best results in transferring maternal immunity to the calf via the colostrum, feed about 5% of calf's bodyweight, 15 minutes after birth (use a nipple bottle). Only use an oesophageal tube if the calf is too weak (e.g. difficult birth) to suckle. To prevent digestive disorders, colostrum should be fed at body temperature. It is not recommended to force the calf's head into an open bucket because this method may also lead to digestive disorders.

As a guideline, feed :

2.25 to 2.5 litres for a Holstein or Brown Swiss calf (birth mass >40kg)

1.25 to 1.5 litres for a Jersey calf (birth mass <30kg)

1.5 to 2 litres for Ayrshires, Guernseys and Dual-Purpose breeds (birth mass 30 to 40kg).

feed again within 8 hours and thereafter 3 times per day at 10% of body mass, for the first 3 days.

Clean the equipment thoroughly after each use to minimise bacterial growth and the risk of transferring pathogens.

Surplus Colostrum

- Can be stored for future use by freezing in flat trays. Storage temperature of –20 oC.
- Defrost in a microwave as needed.
- It is wise to keep some colostrum from older cows as their antibody levels are higher than those of first- calvers.
- Do not store colostrum from cows that have been treated with antibiotics.

Separate Cow and Calf

The longer cow and calf are left together, the stronger the bond that forms between them, making the eventual separation more traumatic for both calf and cow. Calves allowed to nurse take longer to accept bottle- or bucket- feeding. The calf can also ingest dirt from the cow's udder, legs or belly.

Make sure the calf is completely dry before moving it to the calf barn or hutch. Calves can withstand cold temperatures well when their coats are dry.

Identify the calf, record the date of birth, dam's identification, sire and sex of the calf. (Don't rely on memory!) Also record any special notes/details, such as assisted calving, milk fever etc.

Housing

To rear a calf away from its dam and other cows, is to deprive it of its natural form of nutrition, protection, warmth and shelter. For this we substitute a bucket, a building and a stockman! The person/s caring for the calves and overall level of management is infinitely more important than the type of housing used. It is essential to create an environment for the calf which gives the optimal conditions for disease-free growth and comfort.

Some basic considerations to apply to existing housing or to the construction of new housing include :

Site

- Easily accessible.
- Preferably not close to older stock.
- Sheltered, but not surrounded by other buildings.

In South Africa, long walls should face in a North-South direction to avoid excessive heat in the calf-house due to direct sunlight.

Temperature

- Recommended temperature in the calf-house when calves are placed is 15 oC.
- Fit a thermometer to the calf-house and preferably a hygrometer as well. In South Africa, overheating and humidity are a common problem. In the colder areas, protective surrounds can be used for short periods.

Humidity

Should be maintained above 50%, optimally at 70%. Low humidity tends to dry out the respiratory passages and predisposes the calves to infections. High humidity and high temperatures provide an ideal environment for bacterial multiplication.

Goals of Feeding Before Weaning

- To feed for optimum health and strong immunity.
- Encourage the intake of a starter as soon as possible.

- Promote good skeletal development.
- Develop the rumen.
- Prepare the calf for weaning.

Providing for moderate gains and encouraging dry feed consumption before weaning then results in more economical gains, good rumen development and a healthy appetite for dry feed.

Milk Feeding

Milk feeding systems fall into two broad categories:

Ad-lib or high milk intake – calves are given high levels of milk and will grow rapidly up until weaning because milk is the perfect food for young animals. However, if the intake of solid feed has been limited, rumen development will be poor, leading to a check in growth after weaning. This is a waste of the advantage conferred by feeding expensive, highly nutritious milk previously.

Restricted milk intake – calves on restricted milk intakes require nutrients from another source if they are to match the growth rates of calves with ad lib milk intake. The introduction of a high quality calf starter from as early as day 1, will promote growth rates up to and past weaning.

Whichever policy is employed, the objective must be to have calves growing consistently to weaning and beyond, to avoid a post-weaning check in growth.

Colostrum / transition milk should be fed for the first 3 days after birth - feed at 10% of birth mass :

- 2 litres twice daily for a Holstein / Friesland, i.e. 4 litres a day
- 1.5 litres twice daily for a Jersey calf, i.e. 3 litres a day
- 1.5 to 2 litres for the other breeds, i.e. 3 to 4 litres a day.

From day 4, after colostrum, the calf should be offered milk that has the highest nutritional value to allow satisfactory growth for the lowest price. The following factors are important when feeding milk to young calves: -

Type of milk offered

Whole milk - can be fed until weaning. Limited amounts of whole milk, supplemented with a good quality calf starter are an excellent combination for dairy calves.

Extra / unsaleable milk, e.g. extra colostrum, transition milk

Mastitic milk - prevent contact between calves for at least 30 minutes after feeding to prevent infectious agents from being transmitted from calf to calf. Mastitic milk may increase the risk of health problems. Antibiotic residues may lead to the selection of drug-resistant bacteria, making antibiotic treatments less effective over time.

Milk containing antibiotics / anti-inflammatory drugs - as above.

Skim milk - derived from on-farm-processing - is relatively high in protein, but contains less energy and fat soluble vitamins (Vitamin A and Vitamin B) than whole milk because the fat has been removed. Use skim milk only when the calf is eating sufficient calf starter (about 3 weeks old). The calf starter will provide the vitamins and minerals that may be lacking in the skim milk. Feeding skim milk to calves in cold housing / during cold winters is not recommended as the calves need additional energy to protect themselves against low temperatures.

Milk replacer - may be used from 7 days of age. Replacers may contain less fat thus less energy than whole milk on a dry matter basis. Calves fed replacer may gain at a slower rate than those on whole milk.

Amount fed = meal size

8 to 10% of birth mass (as in 6.2 above) until weaning. By limiting milk consumption, calves are encouraged to consume solid feed at an early age.

Frequency of feeding

Preferably two equal meals per day.

Once a day feeding is possible only under strict and skilful management. When the amount of milk needed per day, is offered in one meal, the capacity of the abomasum is exceeded and the excess milk flows back into the rumen where it may cause digestive upset, e.g. bloating. Another problem with once a day feeding is the increased frequency of diarrhoea.

Method of feeding

- Nipple feeding - forces the calf to drink more slowly and reduces the risk of digestive disturbances and diarrhoea. Strict equipment hygiene is imperative.
- Bucket feeding - a calf can be taught how to drink from a bucket within a few days after birth. This technique is easy, rapid and requires less cleaning work.

Milk temperature

- Ideally milk should be fed at body temperature, 3°C.
- Cold milk can be used, but takes energy to heat in the stomach, which could better be used for growth.
- A consistent temperature at **every** feeding and at the same time every day is essential.

A point to note: disease problems during milk rearing can reduce subsequent heifer growth rates and thereby increase the age at first calving.

Hay, Concentrates and Water

Water

- Make water available at the same time as calf starter is offered. It is important to withdraw water for at least half an hour before and half an hour after the milk or replacer is fed, in order to prevent milk going into the rumen, causing digestive upsets.

- The intake of water stimulates the intake of dry feed, which leads to improved gains and reduced incidence of scours.
- Bacteria in the rumen can only survive if they are in a water environment, i.e. that which comes from free water intake. Milk and replacer does not constitute free water - it bypasses the rumen via the oesophageal groove to the abomasum directly.

Solid feedstuffs - developing the rumen

At birth the rumen is small, underdeveloped and sterile. In the first few weeks of life, the size and proportion of the calf stomach compartments go through drastic developments. These changes are affected by diet. At birth, there are no microbes present and the digestive system functions as that of a monogastric (single-stomached) animal. The reticulum and rumen (fermentation vat) makes up some 30% of the stomach compartments, even though it is non-functional. The omasum makes up about 10% and the true stomach or abomasum makes up the remaining 60%. The abomasum is the only stomach fully developed and functional. As a result, only liquid feed (milk) can be utilised effectively by pre-ruminant calves, a few days old. By 12 weeks, the reticulo-rumen often makes up more than two-thirds of total stomach weight. It has then grown in size and function to the point where it is the major part of the stomach system. The rumen is most important to the animal because it acts as a fermentation vessel for microbes to digest complex carbohydrates and high-fibre feeds, which can then be utilised by the animal. Development of the rumen microflora and papilli (finger-like projections developed from the rumen wall) does not happen automatically. Microflora begin to grow as soon as the calf has a regular intake into the rumen of fibre (cellulose) and starch from dry, solid food.

The wall of the rumen of the newborn calf is thin and paperlike, with no visible papilli. Therefore the rumen has little surface area for absorption of nutrients. Young calves, fed primarily liquid diets, do not function as ruminants because they have only one functional stomach, the abomasum. When a calf is fed milk or replacer, closure of the oesophageal groove allows milk to by-pass the reticulo-rumen and flow directly into the abomasum. Solid / dry feeds go straight into the reticulo-rumen, as in older heifers and cows. As liquid feeds are decreased (weaning), the oesophageal groove gradually ceases to function. The rumen microflora become established in the rumen and rumen papilli development begins, allowing for a much greater surface area for absorption of the nutrients produced during rumen fermentation.

The most important objective in early rumen development is to provide the stimulus needed for EPITHELIAL TISSUE (papilli) development. The epithelial layer on the inside of the rumen is the area where nutrients are absorbed from the rumen. It is therefore crucial that as large a surface area as possible is developed here. This can be achieved by feeding a highly palatable calf starter (coarse grain mix or pellet) rather than hay. Rumen fermentation results in the production of various volatile fatty acids (and other components). When grains ferment in the rumen, some of the major end products of digestion are propionic and **butyric acid**. Forages are primarily digested into another volatile fatty acid, acetic acid.

If calves stay on diets consisting of only milk or replacer, the rumen will remain small and underdeveloped. Calves that receive grain from day 3 onwards have larger, more developed rumens. Calves fed only forages will have large rumens, but their papilli development will be limited. WHY? The **butyric acid** (from grains) produced in the rumen by the microbes is absorbed through the rumen walls as a normal part of rumen metabolism. It is this butyric acid which provides energy for the rumen wall hence providing energy for the growth and development of the papilli. The process is one that is self-generating, enabling the early grain-fed calf to have a tremendous amount of rumen development in 3 to 4 weeks.

One can tell that the rumen has become functional when cud chewing is observed, at 2 to 4 months old. Thus, the availability and early consumption of solid food allows rapid rumen development and early weaning. Note that milk and replacer supports the development of AEROBIC microbes, while dry food supports the growth of ANAEROBIC microbes. The type of dry feed will determine the type of microbes and the type of volatile fatty acids produced.

Do not wean a calf until the rumen is functional and capable of supporting its nutritional needs. The result will be rapid and consistent growth after weaning. Without a well-developed rumen at weaning, slow or little growth will be experienced for at least 3 to 5 weeks after weaning.

Calf Starter

A good calf starter should have the following specifications:

18% crude protein with 13 MJ ME/kg energy.

Palatability is very important. Calves tend to prefer a coarse textured or pelleted feed. Offer small amounts frequently - **keep it fresh**. Overfilling the container is wasteful as the feed may go mouldy at the bottom, if not checked twice daily. Any "leftovers" can be fed to older calves.

Avoid contamination from birds or the calves themselves.

Ensure clean fresh water is available - consumption of dry feed is enhanced by increased water consumption.

By 6 weeks of age, the calf should be eating 650 to 750 grams (small breeds) and 1 to 1,25 kg (large breeds) of starter per day.

Hay

Making hay available to a young calf before weaning can significantly reduce energy intake. The fibre requirements of calves can be met without feeding any conventional forms of roughage. The suggested approach is to add highly digestible fibre to the starter in the form of bran or molasses meal. If none of these can be used, adding 5% good quality lucerne hay to the starter should be sufficient. Commercially available rations already include bran and molasses. Commercial complete calf rations should include lucerne meal. With a well-balanced starter, fed ad lib, it should not be necessary to feed any hay until 6 weeks of age. A limited amount - 0.5 kg per day can be offered from 6 weeks of age.

Do not waste good quality roughage on calves pre-weaning.

Weaning

Weaning should take place when a calf is growing well and consuming the starter, as detailed above. Do not wean small or weak calves. When any changes are to be made, make them gradually.

Weaning later than 8 weeks of age is costly because the cost of milk or replacer is more expensive than forage and concentrates. Growth rates remain limited as long as calves are fed a liquid diet. Weight gain increases considerably after weaning, provided the calf is well adapted to a diet of solid food (starter and forages).

- Feeding and management from weaning onwards
- Management goals of feeding post- weaning

Age at First Calving: Once a calf is weaned, the objectives from 3 months until breeding should be:

- to establish a desirable rate of growth
- to feed the most economical sources of protein, energy, vitamins and minerals
- ensure that the heifer does not get fat, thereby having a deleterious effect on subsequent milk production.

Sexual maturity and initiation of oestrus cycles are more dependent on:

- size and weight of the heifer as opposed to the age
- rapid growth - will lower the age at which a heifer can be bred. It is now generally accepted that the ideal age for replacement heifers to calve for the first time is 24 months.

Increasing the age of first calving only leads to a 0.8% (for Frieslands) and 0.6% (for Jerseys) increase in 305 day milk production per month increase in calving age over 24 months. This small benefit in production is lost once the first calving exceeds 27 months.

The three main disadvantages of increasing first calving age to more than 24 months are:

- For every 1-month increase in average first calving age, an extra 2 to 3 heifers are needed per 100 cows. This means that the number of replacements that have to be managed and fed is also increased.
- The entrance of genetically superior animals into the milking herd is delayed.
- The cost of rearing a heifer is increased as the age at first calving increases. The reason being that a heifer having a lower daily gain makes less efficient use of feed because a larger portion of her feed intake is needed for maintenance. Growing heifers so that they calve younger than 24 months, although possible is not desirable as this leads to a reduction in milk yields in the first and even the second lactation. Calving difficulties (dystocia) can also be a problem.

Weight at First Calving

Body weight at first calving has a more significant effect on first lactation performance than age. A useful guideline is that directly after calving, (not including the weight of the calf and placenta) heifers should weigh 80 to 90% of the breed's mature weight.

Growth Rates and Target Weights

In order for a heifer to reach 80 to 90% of its mature body weight by 24 months of age a minimum daily growth rate is required. The first important target weight-for-age after weaning is when the heifer reaches puberty. The onset of puberty is largely a function of body weight, usually reached when the heifer reaches 40 to 50% of mature weight. Research has shown that a very high rate of gain pre-puberty leads to a reduction in the heifer's subsequent lifetime milk production potential. The reason for this has to do with the development of the udder. The udder consists of three areas of milk secretion and storage:

- the teat and udder cistern]
- large and small ducts
- the alveoli

About 70 to 80% of the milk is produced and stored in the alveoli and ducts and it is in this area that rapid pre-puberty growth rates have a very important effect. Growth and development of mammary cells occurs in phases related to reproductive development. These are:

- FOETAL STAGE TO 2 MONTHS OLD - basic udder structure is formed. No alveoli are present. During this period, the mammary cells are growing at the same rate as the rest of the body's cells.

- FROM 2 MONTHS TO PUBERTY - cell growth in the mammary tissue is double that of the rest of the body. At this stage the "fat pad" is laid down in the udder. Ducts are developing, but there are still no alveoli. The ducts need the fat pad for their development.

- AFTER PUBERTY - cell growth in the udder returns to the same rate as the rest of the body. Alveoli formation begins and there is greater differentiation and numbers of cells, especially secretory cells. Milk production potential is directly related to the number of mammary cells, especially those of the alveoli. If pre-puberty growth is not controlled, then the space left available for the formation of the alveoli and small ducts, will be limited by the larger ducts and the large fat pad.

Bibliography

- Berg J, Tymoczko JL, Stryer L (2002). Biochemistry (5th ed.). San Francisco: W.H. Freeman. p. 603. ISBN 0-7167-4684-0.
- Bourdon, RM (2000), Understanding animal breeding, Englewood Cliffs, N.J: Prentice Hall, ISBN 0-13-096449-2
- Cameron, N. D. (1997), Selection indices and prediction of genetic merit in animal breeding, Oxon: CAB International, ISBN 0-85199-169-6
- Corbridge (1995). Phosphorus: An Outline of its Chemistry, Biochemistry, and Technology (5th ed.). Amsterdam: Elsevier. ISBN 0-444-89307-5.
- Dalton, C, Willis, MB (1998), Dalton's Introduction to Practical Animal Breeding, Oxford: Blackwell Science, ISBN 0-632-04947-2
- Fleming G., Guzzoni M. Storia cronologica delle epizoozie dal 1409 av. Cristo sino al 1800, in Gazzetta medico-veterinaria, I-II, Milano 1871-72
- Hall S and Juliet, Clutton Brock. Two hundred years of British farm livestock, Natural History Museum Publications, London 1988
- Hammond K. Gianola, D (1990), Advances in Statistical Methods for Genetic Improvement of Livestock (Advanced Series in Agricultural Sciences), Springer-Verlag Berlin and Heidelberg GmbH & Co. K, ISBN 3-540-50809-0
- Jules, Janick; Noller, Carl H. and Rhyker, Charles L. The Cycles of Plant and Animal Nutrition, in Food and Agriculture, Scientific American Books, San Francisco 1976
- Juliet, Clutton Brock. Horse power: a history of the horse and donkey in human societies, National history Museum publications, London 1992
- Juliet, Clutton Brock. The walking larder. Patterns of domestication, pastoralism and predation, Unwin Hyman, London 1988
- Lippard, S. J. and Berg, J. M. (1994). Principles of Bioinorganic Chemistry. Mill Valley, CA: University Science Books. ISBN 0-935702-73-3.
- Manger, Louis N. A History of the Life Sciences, M. Dekker, New York, Basel 2002
- Massey, JW and Vogt, DW (1993), Heritability and Its Use in Animal Breeding, Department of Animal Sciences, University of Missouri
- Mrode, R. A. (1996), Linear models for the prediction of animal breeding values, Oxon: CAB International, ISBN 0-85198-996-9
- Nelson, D. L.; Cox, M. M. (2000). Lehninger Principles of Biochemistry (3rd ed.). New York: Worth Publishing. ISBN 1-57259-153-6.
- Newman, S, Rothschild, MF (2002), Intellectual Property Rights in Animal Breeding and Genetics, Wallingford, Oxon, UK: CABI Pub, ISBN 0-85199-641-8

Index

www.ingramcontent.com/pod-product-compliance
Ingram Content Group UK Ltd.
Pitfield, Milton Keynes, MK11 3LW, UK
UKHW021328190726
13845UKWH00004B/73

9 789349 964976